D0643385

The Lean Communications Provider

Surviving the Shakeout through Service Management Excellence

Elizabeth K. Adams

Keith J. Willetts

McGraw-Hill

New York San Francisco Washington, D.C. Auckland Bogotá
Caracas Lisbon London Madrid Mexico City Milan
Montreal New Delhi San Juan Singapore
Sydney Tokyo Toronto

McGraw-Hill

A Division of The ***McGraw-Hill*** *Companies*

Library of Congress Cataloging-in-Publication Data

Adams, Elizabeth K.
The lean communications provider : surviving the shakeout through service management excellence / by Elizabeth K. Adams and Keith Willetts.
p. cm.
Includes index.
ISBN 0-07-070306-X (h)
1. Telecommunication—Management. 2. Telecommunication—Customer services. 3. Telecommunication—Deregulation. 4. Organizational effectiveness. I. Willetts, Keith. II. Title.
HE7631.A3 1996
384'.068—dc20 96-14494
CIP

1 2 3 4 5 6 7 8 9 DOC/DOC 9 0 0 9 8 7 6

ISBN 0-07-070306-X

The sponsoring editor of this book was John Wyzalek; the editor was Susan W. Kagey, the executive editor was Lori Flaherty and the production supervisor was Katherine G. Brown. This book was set in ITC Century Light. It was composed in Blue Ridge Summit, PA.

Printed and bound by R. R. Donnelley & Sons Company, Crawfordsville, Indiana.

MH96
070306X

Contents

Part 2 It's All about Integration

Foreword

Today it is generally accepted that in our world of competitive choice, the customer is king. However, it is questionable whether we all understand this to mean the same thing. Often it is all too tempting and too easy to become immersed in and focused on our own particular area of technology without stopping to think if it is really what our customer wants or whether it can be delivered in a form our customer can profitably use or even understand.

Slowly but inexorably, the technologists are coming to realize that they must accommodate the customer, not the other way round. Yet to supply the market, we must first understand what it is and what it requires. Failure to do this only results in the market leaving us behind and going elsewhere.

For many companies, this is a major change in thinking. In telecommunications, historically the market has been led by technology and production. Competition was not really an issue; that was for other industries to worry about. Things were done in a certain way because, according to the technologist, that was the only way—and no one could gainsay them.

Rapid advances in technology, price performance, and global competition no longer permit such indulgences. Today it is our customers who are increasingly setting the agenda by specifying their needs in their terms and challenging us to provide the solution they want rather than one it suits us to provide.

These customer needs, particularly in the arena of international business telecommunications, are growing rapidly in scope, both geographically and technologically. Many customers are beginning to ask that their suppliers provide only one contract and one contact for all their telecommunications requirements.

This development raises a further issue. The delivery of communications services means that service providers are dependent on each other and their suppliers. Only together will we be able to provide efficient solutions for our mutual customers. Separately, we are unlikely to meet their requirements for sophisticated international services—"one contract, one contact"—nor to meet their needs for responsiveness and price. We must learn to work together effectively.

We cannot yet clearly define the future shape of the telecommunications service provider market. At the end of a long period of shakeout, there might be only three or four global players. However, this shakeout is likely to take a long time, and the number of winners is uncertain. What is more certain is that the winners will be those companies that are best able to respond to, and deliver, what customers require—companies that work hard to make it easy for their customers to do business with them.

To achieve this objective will mean, in some cases, developing partnerships, while in other cases it will mean working with competitors. Certainly it will mean finding innovative ways of delivering new services faster, at less cost, and with greater consistency of quality.

Already the traditional telecommunications players can see strong new competitors eyeing their markets. These new entrants might choose to work with traditional providers. However, they might just as easily decide to work with new alternative service providers, such as cable companies. Whatever their direction, they cannot be ignored. They are light on their feet and used to rapid change. The traditional players must learn to be the same.

In responding to change, though, we cannot forget our own heritage. Plain domestic telephony is still the core business that provides the income and the infrastructure on which we can build our more exotic offerings. Nevertheless, we risk being restricted by this very infrastructure if certain key changes are not made.

We await the worldwide liberalization of domestic telecommunications regulations with anticipation. Changes in technology are also needed. The growth in mobile, satellite, and messaging services, and the increased global use of the Internet have already had an enormous impact, and they will continue to shape our market.

Most important, however, change is also needed in the way our industry does business across borders that once defined territories but which today often represent anachronistic bottlenecks. Cooperation is vital if the market is to develop as it should.

The Chinese proverb "May you live in interesting times" can either be taken as an expression of hope and anticipation or, as its author intended, as a curse. Our industry is entering "interesting times," and those companies that survive the next few years will have truly understood the need for flexibility and speedy response.

This book addresses issues that lie at the heart of what it takes to succeed in the new world of communications. The authors take a long-term view but also map a path to help us get from where we are today to a healthier future. Service providers will do well to heed their message.

Viesturs Vucins
President and CEO of Uniworld

Uniworld is a company in formation as a joint venture between AT&T and Unisource, which in turn is owned by KPN, Telia, the Swiss PTT, and, pending regulatory approval, Telefónica de España.

Acknowledgments

Most people never undertake to write a book, and we can only believe that those who do find it an experience somewhat more difficult and certainly more time-consuming than they expected. We certainly did, and our families unfortunately bore the brunt of our efforts. For that reason, we would like to thank our partners, Bill Adams and Kirianne Willetts, and our children for their patience and support through some very trying times.

In addition, we thank Bruce Murrill for his thorough technical review, Wendy Shorrock for injecting a more realistic perspective into some of our loftier visions, Rod Matthews for helping us properly position the book, Bill Adams for his comments and valuable references to current accepted quality principles, and Alison Hilton for turning illegible handwriting into readable text. Others who willingly shared their stories are named within the book, and we thank them for enriching our thoughts.

Preface

It is doubtful that Judge Harold Greene and Prime Minister Margaret Thatcher realized, in overseeing the introduction of competition in the U.S. and U.K. communications service industries, that their actions would ignite a firestorm that would rage across the entire global marketplace. Now, more than a decade later, it is clear that nothing short of a complete reinvention of this dynamic industry is underway. Both of us have been firsthand witnesses to many exciting changes on both sides of the Atlantic and have learned from our associates how similar changes are affecting service providers and suppliers in different cultures and regulatory environments.

All over the world, communications companies are facing radical changes to their once-stable and protected domain. The effects of deregulation, privatization, and consequent unbridled competition are causing service providers and their suppliers to fundamentally rethink the way they do business. This industry is undergoing a revolution in the way it is shaped, organized, and operated. And revolutions tend to be bloody, with victors taking the spoils, empires toppling, and new structures emerging. Have no doubt, the revolution in communications will be harsh. New entrants are crying "take no prisoners" in their battle for market share. Nor are the established players sitting idly by watching their markets decline.

Who will survive the inevitable shakeout? The battle lines are becoming clearer as the protagonists (the service providers), aided by their arms dealers (the suppliers), lock in mortal combat.

Already, the provision of communications services in many countries has evolved from a moribund state bureaucracy to a highly competitive and fast-changing marketplace. Viewed from the perspective of the customer, many of the changes are starkly obvious: Real prices have fallen dramatically, while at the same time, service provider responsiveness, levels of customer service and performance, and speed of innovation have changed beyond recognition. Already, the outlines of a totally reshaped industry are starting to take form. Alliances are being created, lower barriers to entry are allowing new companies to enter markets on an almost daily basis, and increasing convergence is creating new competitors from unlikely sources.

However, many of the biggest changes are taking place "below the waterline"—inside this $750 billion industry, where they are not always obvious to the public and press. And just like the *Titanic*, it is precisely here that many of the world's communications companies are most vulnerable.

To that end, this book is mainly an insider's view of where companies are facing threats and what they can do about them. More specifically, it examines the current state of providers' service delivery operations and the overriding need to simultaneously cut costs, improve service quality, and reduce time to market of new service offerings. It looks at the crucial need for service providers to reengineer towards an integrated, automated service delivery platform, replacing their existing fragmented hodgepodge of systems and processes and making them capable of delivering profit in the new competitive order. It is, in short, about what it takes to become a "lean service provider" excelling at service management—a fundamental requirement if a company is to survive in this industry.

Success in the new communications service business depends on the ability to emulate manufacturing industries by automating end-to-end flow-through of service management and network management processes. Failure to achieve this kind of lean provider status spells potential disaster for some companies as the shakeout that will follow global liberalization of markets eliminates the weak and the lame. Those companies that are saddled with high costs, poor levels of customer service, and slow reaction times will lose business to those that have learned to operate more efficiently and effectively.

And, while success requires that lean providers make competitive moves, it also depends on being able to effectively collaborate with suppliers, allied providers, and even competitors. The communications industry has a history of cooperating in matters of network connectivity (although anyone trying to use a cellular phone in the U.S. might argue that the newcomers have some lessons to learn), but they have not been faced with the need to link their management processes and systems until recently. As with all new things, the urge to resist change in this crucial area is strong, and it doesn't help matters that the management infrastructure is massive, outdated, hard to change, and different in every company.

The revolution in the communications industry will keep the case-study writers of Harvard and MIT busy for decades, and this book is not intended to be a wide-ranging review of all of the business impacts of that restructuring. It does, however, go to the heart of the inside of the communications industry, where managers are faced with finding radical new approaches to delivering services to survive the competitive jungle.

This book is intended for managers who see the relationship between survival and the health of service management systems and processes but who are looking for help on how to go about making needed changes. It is for managers who see their competitors moving faster and at lower cost and who are discovering that it's not enough just to downsize. It is for managers who see the day looming when their company will be liberalized and who need to know how to prepare. It is also for suppliers who need to know how they can best respond to their customers and help them succeed.

This book can't and won't address all of the problems in delivering service management excellence to customers. Many of those are areas of competitive differentiation in which companies that find the answer faster will reap the rewards. This book does, however, explain what service management is all about for those not accustomed to the operations side of the business. It highlights places where service providers need to take action in common with their brethren, either to ensure that the industry as a whole will deliver what customers need or to influence the cost and functionality of suppliers' equipment.

This book is not a technical treatise, although it covers some of the most fundamental technical strategies that support effective service management—strategies that executives in this industry should strive to understand. Neither is it a conceptual treatise. It takes as its focus the issues involved in trying to implement change through process flow-through improvement and integration of service management systems, and it discusses the practical realities associated with implementation.

On a personal note, this book is, for us, a way to summarize where we are in our thinking and to put in one place all of the many different themes we have developed separately or together during our years with the Network Management Forum (NMF), an industry organization dedicated to finding industry solutions to common service management problems. If you have read some of the NMF's more strategic publications, you may find a few sections in this book that look familiar or whose theme is similar to what you've read before. As authors of the original text, this was a good opportunity to sift through our earlier efforts, to save and amplify ideas that are still sound, while discarding those that have gone by the wayside.

In writing this book, we have had the help and support of our families, friends, and colleagues, for which we are very grateful.

Elizabeth K. Adams
Keith J. Willetts

Part

1

When Service Is Your Business

Providing the ability for customers to communicate—in voice, data, or pictures—is probably the most intangible "product" any company might sell. Customers see the benefits of a communications service, but they can't touch it. The equipment customers use to access these services (telephones, private switches, and computers) is usually purchased from a different company, or at least carries a different logo, so the customer doesn't see tangible reminders of the service provider each time they make a telephone call or access the Internet. Unlike the airline service industry, where customers are surrounded by airline personnel at the check-in desk and on the plane, or the parcel delivery business, where at least a regular relationship is formed with the person making pick-ups or deliveries, communications service providers are nearly invisible.

Most of the time, the customer doesn't think about his or her experience in using a communications service—we don't think "Wow, that was a really good call!" Like water and electricity supplies, we only notice when things go wrong or we want something changed. And that is the true point of delivery of the customer experience. Several years of good-quality communication service can be shattered in a few minutes if the customer care processes are poor.

When service is your business and that service is both invisible and a part of the fabric of the customer's life or business, success can only be achieved by doing three things:

- *Providing excellent value (keeping prices competitive)*
- *Providing excellent quality (meeting or exceeding customers' expectations, both in terms of the underlying communications capability and in any customer interaction)*
- *Continually offering new ways in which services can make life better for the customer*

These challenges can be boiled down to three basic objectives that must be continuously met: reduce cost, improve the quality (and the perception of quality) of the service, and reduce time-to-market of new services. Anyone who has tried to do even one of these has found just how challenging it can be. Doing all three simultaneously requires a fine balancing of objectives that oppose each other. We call this activity, this balancing of opposing objectives, service management.

For any company operating as a service business, finding this balance is key, and it is extremely difficult even in a stable environment. But the environment faced by communications service providers is anything but stable, and service providers must achieve these three objectives while also facing the pressures that come any time an industry moves from a regulated to a market-driven structure: meeting shareholder expectations, establishing an identity (differentiating itself from the competition), and creating customer loyalty.

For all but the new entrants, service providers have another cross to bear. Their internal processes and systems are antiquated and inflexible, and that is a major bane to service management excellence. They are unaccustomed (and usually unable) to do business effectively with external organizations and therefore fail to respond as their traditional vertically integrated industry becomes increasingly segmented.

For these companies to excel at service management calls for radical new ways of looking at their business, including both internal processes and external relationships. This book is aimed at helping service providers see just how differently they need to think about how they do business and how they can move from their existing environment to one that will deliver service management excellence, allowing them to become lean service providers.

In this first part of the book, we explore the industry itself in more detail—identifying the players (both old and new) and the state of liberalization worldwide. We then take a look at what service management means in this industry and review the basic approach being taken by leading service providers to address the many different types of management systems and processes that are needed. Finally, we highlight those areas in which service providers are likely to need to create external linkages with their customers, their suppliers, and each other.

This part sets the stage and describes the problem the industry is trying to solve. The parts that follow it explore, in turn, the concept of integration as applied to processes and systems, what is available from industry groups to help service providers achieve integration of their service management systems and processes, and steps a company can take to get from where they are today to where they need to be.

Chapter

1

The Sleeping Giant Awakens

In the United States, it has now been more than a decade since the breakup of the Bell system into long distance and local exchange carriers created a partially competitive environment. The long distance (or interexchange) carriers have had to engage in significant cost-cutting measures in order to keep pace with the price wars they have waged to keep or gain market share. The local exchange companies, on the other hand, continued to operate in a quasi-monopoly environment until 1996, when legislation to remove the protections afforded the local exchange carriers (and lift the prohibitions that prevented interexchange carriers from openly competing for local business) was finally approved.

The U.K. market has been liberalized for nearly the same length of time. Initially there was a period of several years of a duopoly between BT and Mercury, after which the market opened further to allow a wide range of competitors to provide communications services. The market has become fiercely competitive, and at the time of writing, more than 150 companies had a telecommunications license in the United Kingdom. Unlike in the United States, the regulatory environment in the United Kingdom has not imposed a local-interexchange split, and it is possible for any company to compete in any part of the domestic communications industry. Major service providers compete in multiple market segments (local, interexchange, international, and mobile services), and, at the same time, new entrants such as foreign carriers, power companies, cable TV operators, and others are offering selected services, including basic transport, access, or cellular-based services.

In other European markets, liberalization is more patchy, with some countries, such as Sweden and Finland, operating in a U.K.-style, highly-competitive environment. Equally, there are examples within Europe of countries, such as France and Belgium, where movement toward liberalization is slower, and where a state-owned monopoly structure may exist until 1998, when European Union agreements demand that services and infrastructure should be fully open. Although some EU members have extensions to 2003, this move will eventually create an EU-wide communications market larger than that of the United States.

Japan already supports a bifurcated environment, with NTT providing domestic services, and service carriers (such as KDD) providing international services. In addition, within the domestic market, many small carriers offer long-distance services in competition with NTT. Action is expected during 1996 that will modify the current environment by restructuring NTT in some fashion, possibly following the U.S. model.

Elsewhere in the world, examples abound of highly competitive environments (such as New Zealand, the Philippines, and Chile), as well as the fast growth of wireless networks in developing economies. The attraction of these growing markets has sparked considerable investment by established operators whose home markets' growth is more conservative. According to *Business Week*[1], in Chile alone, the opening of the telecommunications market caused the number of carriers to more than quadruple in one year, from two to nine, with all competing feverishly for a cut of Chile's 14% annual growth in international calling minutes. Among the players: U.S. RBOC's BellSouth, Bell Atlantic, and SBC Communications, Italy's STET, Telefónica de España, and Samsung Group. Figure 1.1 shows how quickly developing countries are moving to a liberalized, private telecommunications industry (Arnst 1995).

Although there is no single model for how countries have moved or are moving from a monopolistic communications environment to a competitive one, one thing is clear: Competition of one sort or another will happen to virtually every existing communications provider. And if other industries are any example, the industry shakeout that follows the advent of competition may well mean the disappearance of existing providers that are unable to keep up with those who are more efficient and effective.

Competition: Who and Where?

Domestic revenues from telecommunications are often so large that looking at relative turnover between the incumbent player and newer competitors often hides the very significant losses of market share in specific markets. There are various methods of segmenting the market, but for illustration it

[1]Arnst, Catherine. 1995. The Last Frontier. Business Week. September 18:99.

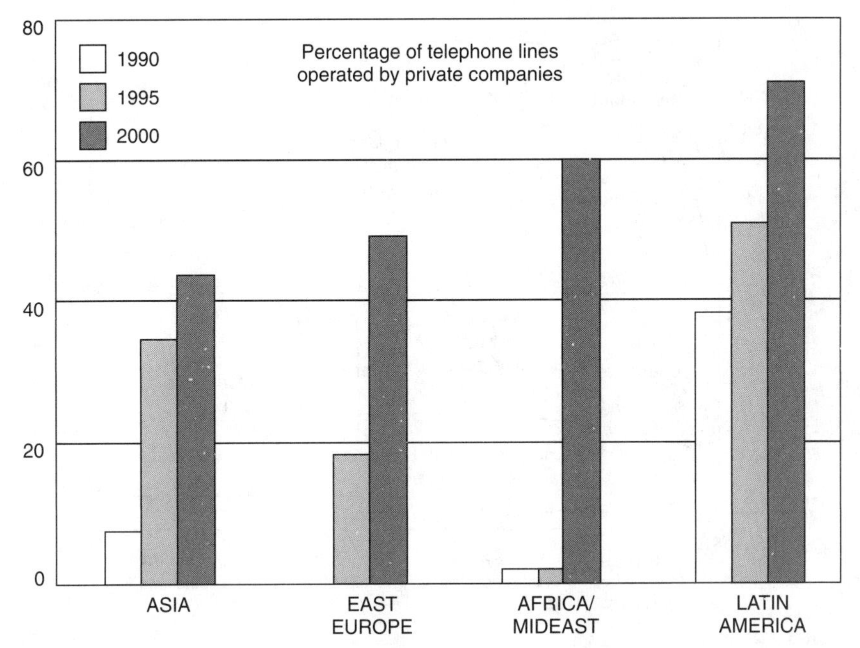

Figure 1.1 The movement toward liberalization. *(Adapted from Business Week)*

is useful to segment the market into international, domestic long distance, and local services. Many regulators differentiate between value-added and basic voice services, so it is also useful to view the market on a service as well as geographic basis. Competitors will appear in all of these segments wherever legislation allows and, as we will see in the case of some international services, regardless of the state of local legislation. The competition itself also comes in a variety of types, described in the following paragraphs and with their geographical and service segmentation shown in Figure 1.2.

New fixed-network operators

New licenses for fixed networks are being granted in an increasing number of countries. These players, including cable TV companies, consortia of utilities companies, banks, and foreign operators, are springing up to challenge the established order. They may focus initially on a market niche to build revenues that will finance further expansion. International services to major corporations are the usual entry point, because these represent the most profitable sectors at the least investment. Domestic long-distance services are the next most attractive sector in most markets, via interconnect to the major player or radio-based bypass to provide local services. Success isn't guaranteed, but more often than not, even small new operators can have a dramatic impact on the incumbent

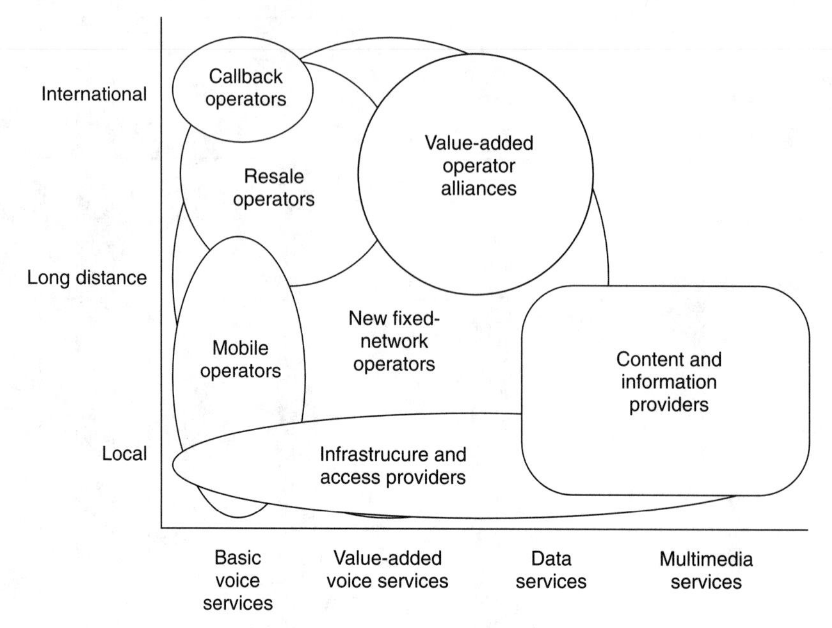

Figure 1.2 Service segmentation.

operator by skimming the high-margin cream through price competition and service innovations.

Mobile operators

Mobile services, mainly using cellular technology at this stage, are growing rapidly in many markets. These will soon be supplemented by low and medium earth-orbit satellite technology (LEO and MEO), which is hovering in the wings. Where existing infrastructure is poor, mobile services represent a way of quickly modernizing basic voice services and are thus favored by governments and regulators wishing to overhaul a country's capabilities. Growth rates can be phenomenal, and traffic is often incremental to fixed network services rather than substitutional. Nevertheless, cellular services, LEOs, and local radio distribution represent a very significant competitive threat to conventional fixed network operators.

Callback operators

Callback companies, often operating in the margins of legality (but very profitably), act as arbitrage speculators to take advantage of differences in prices of international basic voice services. There are a number of methods of callback. In most cases, the customer is provided with a calling card and gains access to the callback equipment via a toll-free number. The customer enters his or her account number and PIN, plus the telephone number that he or she is currently using. The call is then cleared and the callback equipment establishes a call back to the customer and allows international dialing at rates pertaining to the country in which the callback equipment is located—often very much lower than the rates permitted in the country where the customer is located. To give an example, it is currently cheaper to route a call from Korea to Japan via the United States using callback than directly at Korea-to-Japan retail prices.

Some arrangements can be much more sophisticated. In every case, the effect is the same. The incumbent operator in the country from which the customer's call is originated loses market share for highly profitable international calls to the callback company. Although various moves are being made to outlaw such practices and to provide technical solutions to bar such calls, the only certain way to protect market share against callback companies is to offer international rates that are at least as competitive as those offered by the callback companies. Clearly this creates significant downward pressure on prices—even in heavily regulated monopolies.

Resale operators

Resellers are being authorized in many countries and again put pressure on international and long-distance pricing. There are two primary forms of resale. One-ended resale allows connection between a private network and a public network, where typically traffic from a customer's PBX is permitted to "break out" onto a public network, which can be in a distant country. As with the callback companies, resellers can take advantage of the different rates between countries to "hub" via a low-cost country in order to offer customers much lower international rates than those offered in the country where the PBX is located. International simple resale goes one step further and allows connection to the public network at both ends of a private circuit. At present this is authorized between a small but growing number of countries, and it is creating significant downward pressure on prices.

Resellers, along with callback operators, represent a fast-growing segment of the industry. Often not from a conventional communications background, owners of these types of companies inject new ideas and thinking and may present a formidable challenge to established players. They also minimize their network investments, leasing transmission and switching capabilities where economically sensible. Today, many of these companies compete simply on price rather than on the more complex service management mix that is the subject of this book. Those that survive the start-up phase will, however, need to turn to a service management approach as price differentials with their key competitors narrow.

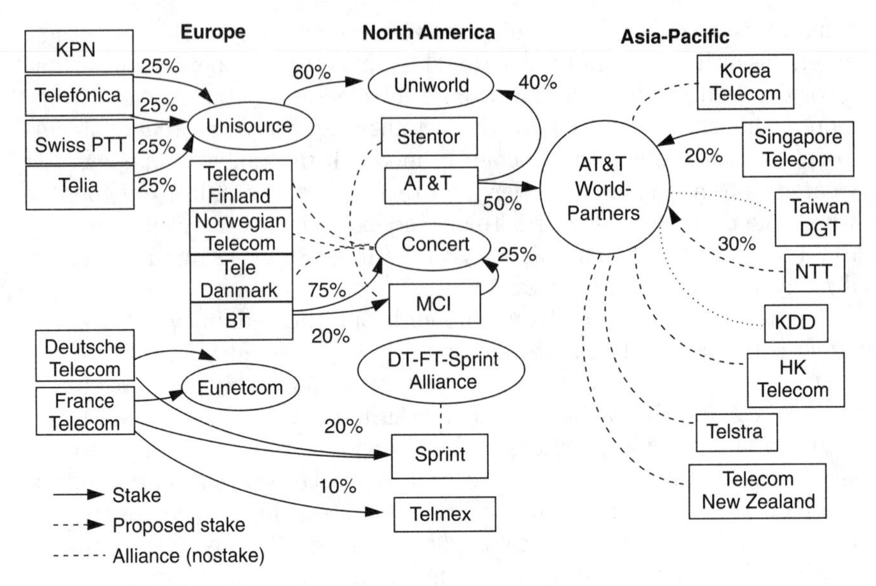

Figure 1.3 Mergers, acquisitions, and alliances. *Analysys, Ltd.*

Value-added operator alliances

The fifth competitor group causing change in the marketplace comprises the emerging value-added global service providers, which often take the form of consortia of multiple service providers. Several major groupings have emerged, such as the BT/MCI Concert alliance, Deutsche Telekom/France Telecom/Sprint's Phoenix alliance, AT&T's WorldPartners, and the Unisource alliance of Telia, KPN, Telefónica, and Swiss Telecom, as shown in Figure 1.3.[2]

Today's alliances are in various stages of evolution, but some, including Concert and Unisource, have launched broad portfolios of value-added voice and data services available in many countries around the world. This approach represents a possible model for global restructuring of the telecommunications industry, especially in the area of international services. Currently these services are controlled by the International Telecommunications Union (ITU) on behalf of the world's operators, dating from a time when national operators were all "friends." Now that these operators are becoming fiercely competitive, the alliance model—groups of collaborating operators—may supplant the traditional approach to international communications.

Alliances represent a significant challenge to the delivery of service management excellence, because customers are demanding seamless, end-to-

[2]Analysys, Ltd. 1994. Strategic Alliances between Telecom Operators. Analysys, Ltd.: Cambridge, UK.

end global service, which in reality may be delivered over multiple partner infrastructures. The sophistication of these global services is such that traffic from country A to country B may be billed in country C, ordered in country D, and have customer service points in all locations. This demands complex many-to-many relationships between the multiple service management systems.

One thing seems certain. Although the alliances are focusing today on value-added services, they can be expected to broaden their horizons toward basic services and in-country services as legislation permits.

Infrastructure and access providers

Liberalization is now allowing competition at the infrastructure level in an increasing number of countries. Companies with physical access to customers and markets, such as railways, electricity companies, pipeline operators, etc., have a valuable resource known as a way-leave or right-of-way. They are utilizing their physical assets to bypass the incumbent operators' local or long distance facilities.

For example, electric companies have perfected a technique of wrapping fiber-optic cable along their distribution networks, allowing rapid and low-cost alternative networks to exist. In Europe, Hermes, a consortium including railway companies, is building a pan-European, high-capacity fiber-optic network, linking all of the major European Union states. In London, an access provider has turned a defunct Victorian steam-pipe network into an advanced digital access network. In the United States, investment in microwave and radio bypass have complemented fixed-access methods.

Some of these access providers are restricting their scope to providing wholesale services to service providers. Some, however, are competing directly with existing providers and offer a range of voice and data services, often over brand-new, easy-to-manage, high-quality synchronous digital hierarchy (SDH) fiber networks.

At a different level, new companies are springing up to offer access to online services and networks such as the Internet. Although these companies rarely provide physical access (they use leased dial-up or packet services) they do represent another example of competitive service providers that could not have been predicted a few short years ago.

Multimedia operators

New technologies such as ATM, wireless, and broadband, plus multimedia services including video-on-demand, home shopping, electronic commerce, and online information services, have the potential to rewrite the rules of the traditional "telecom" sector. This is because competition may emerge from different industries altogether rather than between traditional telecommunications players. The convergence of information technologies is allowing companies in the cable TV, publishing, online services, and computing industries to set their sights on telecommunications revenues.

In the United Kingdom, cable TV companies can now offer basic voice services via their networks, as well as broadcast services and even video-on-demand. The United States may well move in the same direction. So when taking a view of competition, it is necessary to "think out of the box" and see the competition that might come in through the back door as well as the front.

One example is the rapid growth of the Internet over the past few years, providing ubiquitous and cheap electronic-mail (e-mail) services and a wide range of information services. The Internet has caught most of the telecom industry napping and is growing at such a rate that its power to reshape the information and telecom industries is vast. Although better suited to asynchronous traffic (i.e., not real-time) it can, and does, deliver voice and video information as well as data traffic. While this might not replace voice telephony, which demands a synchronous network, we could see the Internet displace facsimile in the same way that fax displaced the telex only a few years ago. And we could see a huge rise in the amount of voice or video messaging over the "Net" as more and more PCs are built with voice-processing capabilities.

As another example, the development of information networks such as the Microsoft Network could also radically alter the current telecommunications landscape. With around 80 million personal computers potentially equipped with Microsoft networking capabilities, it is quite possible for the PC to become a local communications controller offering data, voice, and video facilities with advanced features.

Thus, on multiple fronts, competition is emerging in domestic and international markets. No operator can be immune to these changes because the international communications marketplace is so highly integrated. A particular government may take the view that it will either not liberalize its market or adopt a very slow pace of change. But as with callback, competition is becoming borderless. No country or service provider can be an island, peacefully immune from the revolution that is taking place. The movement towards a fully competitive global market is inexorable—the only question is when rather than if markets will open.

More and more governments are beginning to experience these realities. And they can't ignore the behavior of major corporations, which count access to advanced communications and supplier choice as important issues when deciding where to locate new offices or factories. If liberalizing the market can bring jobs and investment into a country, and selling a state monopoly to private investors can bring in foreign exchange, the political attractions are hard to resist.

The Market Outlook

Robert Allen, chairman of AT&T, has said that anyone with a five-year plan in the international communications market clearly does not understand what is going on. To a large extent he is right. It is very difficult to predict the outcome of unleashing competitive forces in a marketplace dominated for so long by bureaucratic, state-controlled monopolies. It is as though a sleeping giant has been awakened after many years of slumber. Unfortunately, the giant has awakened with a headache and is unpredictable and uncontrollable.

Although we can't be certain of the outcome, we can be reasonably clear about a number of trends.

- First, prices in competitive communications markets will continue to fall for the foreseeable future. Improvements in the underlying network technology are allowing new entrants, with new networks, considerable cost advantages that can be turned into pricing advantages to gain market share from the incumbent, unless the major player moves quickly to reduce its cost base.
- Second, levels of customer service and service quality will continue to rise. Customers focus on value from their suppliers, a judgment that takes into account both price and quality. New entrants can make considerable progress in gaining market share where the incumbent treats its customers badly, takes a long time to fulfill service orders or solve problems, or fails to meet service performance requirements. In much the same way that price is subject to borderless competition, customer service and network quality are also compared across national borders. At the very least, customers with any international experience—particularly large multinationals and international travelers—will benchmark service quality against the best they receive somewhere in the world, which may not be the local provider.
- Third, the rate of product innovation and introduction will continue to accelerate rapidly. For decades, the telecommunications world was very conservative, and new services were slow to evolve. Often, many years of technical standardization preceded even early product development, with additional time to reach full national or international service availability. Today, customer demand for advanced communications is growing at such a rate that an evolutionary approach leads to services that are already out of date in the mind of the customer by the time they are launched. Smart competitors will seek to shorten product introduction intervals as an area of strong competitive advantage. For an incumbent operator with legacy processes, systems, and technologies, this can become a nightmare leading to short-term fixes of overlay networks and even

overlays on overlays. While solving some of the problems of product introduction, this usually adds to the challenge of delivering excellent customer service at the right price.

- Fourth, we can expect to see aggregation and shakeout in the market place. Communications equipment suppliers have already been forced to seek wider global markets as local procurement on a national-favored basis has diminished. The last few years have seen a wave of takeovers and mergers among these companies, and it is likely that this trend will extend to the service providers themselves. Inward investment in small- to medium-sized operators by large players is increasingly frequent as state-owned monopolies go private, and this must be viewed as a precursor to a more fundamental rationalization of the service provider industry. It is not inconceivable that some service providers who fail to manage the transition from monopoly-protected status to the rigors of the open market will fail to adjust and, like Pan Am and Eastern Airlines, simply cease to exist as independent organizations.

  The global alliances will increase in strength, in the breadth of their portfolios, and in the scope of their customer base to emerge as global supercarriers. Despite much hype about moving into the value-added arena, telecommunications remains largely a commodity product, and commodity markets need volume and scale in order to maximize profits. Historically, margins in the telecommunications industry have been very good, often derived on a "cost-plus" basis. In a free and competitive market, margins may well be much slimmer, and communications "groups" will need to grow to maximize their volume in order to unlock economies of scale.

- Fifth, technology advances will continue to bring down the costs of entry into the communications marketplace. Traditionally, financial barriers to entry have been enormous, requiring vast investment in cable, duct, and switching before even one call could be made. New technologies such as cellular are allowing much lower cost of access to these markets. The separation of networking from services also allows very low entry costs for newer service providers. Many entrants have no network infrastructure at all, leasing this capability from others as required. Lower entry costs always ensure that competitors will spring up quickly to exploit market niches or failings of the existing players. As we said before, these competitors may well not come from a telecommunications background at all. Major software companies, such as Microsoft, are entering the service provider market without ownership of the physical networking underneath their information services.

These trends have major implications for service providers new and old alike, and although Robert Allen might be correct in implying that the communications business is moving too fast to make detailed long-range plans, every service provider that wants to be in business in five years—whether a new entrant or an existing operator—must be aware of these underlying trends and take appropriate action just to survive, if not to grow and prosper.

How Are Existing Service Providers Responding?

Faced with the challenges outlined previously, existing service providers are in various states of evolution in their responses. Perhaps a new trivia game might emerge in trying to put names to companies in each of these states. We have our own views, but as people who make their livelihood in this industry, we believe that discretion is the better part of valor and will keep them to ourselves!

The headless-chicken phase

The "headless-chicken" stage usually occurs well before any deregulation or privatization activities have taken place. This is when a government is weighing up the options, and considerable public debate is taking place. This is a very difficult period for a service provider, since it introduces tremendous uncertainty, making it impossible for senior managers to focus their planning on any particular eventuality. The prognostications of politicians, evening news commentators, investment analysts, and fortune-tellers are equally likely to be true or false and equally likely to guide employees as they try to anticipate the future. This phase leads to much abortive planning work and multiple, rapid reorganizations as managers try to get their teams into shape in the time available before competition is unleashed. If you have experienced this phase, as both of us have, you'll recognize that it's marginally better than what usually follows.

The blind-panic phase

During the "blind-panic" phase, the ground rules for liberalization have been set and early competition is underway. During this phase a frenetic amount of work must take place both at a government level and within the service provider itself—bearing in mind of course that the vast majority of monopoly service providers are government departments in their own right. An interesting phenomenon of this phase tends to be an inability to correctly assess the magnitude of any commercial threat, so that a company either overreacts to even minor competitive moves or completely misses real threats.

New entrants can make their greatest impact during this phase since the incumbent probably has yet to master the skills of sector-based marketing. An entrant can penetrate a niche sector before it shows up on the main player's radar. As an example, a major provider was convinced following deregulation that a new entrant would go after its lucrative domestic long-distance revenues from residential customers. The incumbent aggressively attacked the anticipated problem through price cuts and service improvements, all the while missing the fact that their competitor was cherry-picking high-volume business users from major metropolitan areas. By the time the incumbent saw the trend, its customer base of data services had all but evaporated.

A more obscure threat can come from the rise of intercity "carriers' carriers"—companies that make use of power company or railway rights-of-way to lay fiber routes. Initially, service providers might not view these infrastructure providers as direct competition since they generally supply wholesale rather than retail services. But alternative infrastructure providers fuel the rise of competitive service providers by substantially lowering the costs of entry into the market. In turn, a glut of infrastructure alters the long-run economics of networks that need to meet certain utilization levels to produce a return on investment. This could spell bad news for the new player who fails to achieve critical mass or for the incumbent whose high maintenance costs and functionally limited copper networks may be outpaced by the newcomers' more modern technology.

Some service providers have viewed this phase as the "Phony War," a period in which competition doesn't seem to spell disaster. They think they know what a competitive market is all about and fail to make the changes required to compete with companies that have moved to the next level.

The self-delusion phase

In the "self-delusion" phase, the incumbent operator has been deregulated or transformed from a state-owned to a private company, and competition has really started to bite. At this point, a service provider starts to understand the magnitude of change that competition will bring, and the three imperatives we have examined earlier (reducing cost, improving service, and shortening time to market) begin to take hold, although at this stage they are not often seen as three parts of a whole that must be kept in balance.

More typically, the accountants drive the cost-reduction targets (often instituting across-the-board cuts), the operations people drive customer service improvements (usually arguing for more people to take better care of customers and virtually giving away potential revenues in the name of good service), and an explosion of "really good ideas" for new services threatens to surpass the company's ability to ensure that each new undertaking is done well.

Luckily for most incumbent operators, there is usually plenty of fat that can be removed to make a fairly immediate effect on the cost base, and retraining and refocusing of employees can bring about dramatic improvements in customer service. Restructuring, delayering and countless other buzzwords mark this phase. From within the service provider, it's a bit like watching an ex-smoker join a campaign to ban all smoking. Years of slow decision-making give way to a "ready, fire, aim" mentality, and aggressive young MBAs with no experience in communications may become the example to follow.

Trimming the fat and learning to think in terms of customers rather than subscribers can have an enormous impact on the level of services offered and the cost base of the corporation. But it is a finite source of improvement and cannot be endlessly exploited in the race to continuously raise the level of service management excellence. In this phase, downsizing is an easy way to get short-term results, and many companies take cost and labor out without thinking about how to create long-term value. It is at this point that the game will either be won or lost, since improvement from this phase demands the vision to see that things can be done better only if done very differently. Currently, only a small number of operators are starting to develop toward the next phase.

The work smarter, not just faster phase

During the "work smarter, not just faster" phase, senior management realizes that the only way to truly compete is to radically change the way the organization works and that the old bureaucracy can't afford to lose any more weight and still survive. Reengineering and reinventing become the "in" terms as companies go back to the drawing board to design the organization and its key processes from scratch. Reengineering looks for step changes in operating economies and customer service by doing things in a totally different way. The "fathers" of reengineering—Michael Hammer and James Champy in their seminal book *Reengineering the Corporation*[3]—place great emphasis on end-to-end process redesign, built not around a model of work based on people, but one based on advanced information technology.

Here, we get to the core of the subject of this book. How can service providers, both newly established and existing players, continuously raise their game in delivering service excellence? The answer lies in the integration of their support and management systems, both internally to the corporation and externally, as links are created to their customers and cooperating service providers. This allows technology to substitute for people, freeing human skills to be deployed in serving the customer or developing creative marketing campaigns to grow overall market value. You will see the term *process flow-through* used many times in this book as we expose our preoccupation with delivering service excellence through reengineered, end-to-end process flow-through.

---

[3]Hammer, Michael, and James Champy. 1993. Reengineering the Corporation. New York: HarperCollins.

Responding to competitive threats has implications for every part of the service provider's business. In a way, reengineering an existing corporation is quite similar to the task of building a company from the ground up, with many of the same considerations needing to be made. These include:

1. Articulating a vision for the corporation.
2. Developing and executing a market-driven product strategy.
3. Executing a personnel strategy that brings needed skills at lower overall cost.
4. Organizing to quickly respond to globally distributed opportunities.
5. Changing financial practices to reflect a competitive, rather than monopoly, environment.
6. Reengineering the network technology to increase flexibility and lower costs.
7. Reengineering the operational processes and systems to improve service, reduce costs, and speed time to market.

This book is about item 7. It starts from the premise that tomorrow's communications companies will be long on integrated, sophisticated support and management systems and short on operations people—the opposite of most service providers today.

Automated process flow-through sounds complex but in reality is very simple; how do you organize a process so that the number of steps involving costly human intervention can be minimized? The "one-touch" process terminology was introduced by GTE to describe the ideal—take an order at the front desk, and dial tone appears in the customer's home without manual intervention. GTE has gone even further to eliminate the front desk. With its "no-touch" provisioning, a customer moving into a new house simply plugs a phone into the jack, picks it up, and is automatically guided through a series of touch-tone steps to establish new service. One-touch or no-touch processes can transform the cost base and simultaneously improve customer service—a neat trick if you can do it.

Getting to a point where internal and external processes are highly automated and integrated is critical to becoming what we call a lean provider—one who can sustain profitability because of a low-cost, high-quality operational platform.

Chapter

2

Balancing the Three-Legged Stool: Service Management Excellence

It's easy to improve customer service and quality when cost is not a factor—just hire people, as service providers regularly did when their prices and rates of return were cost-based. It's easy to reduce cost when quality doesn't matter—just fire people, as service providers have done when racing for market share through price wars. It's easy to shove new services out the door if neither cost nor quality are factors—just "tweak" the technology and put in an overlay billing system, as service providers continue to do to keep up with new players in the value-added service business.

But as anyone knows who has sat on a three-legged stool, knock out one leg and you will fall over! Service providers who fail to keep these three factors in balance will eventually fall out of the industry. The only way to achieve service management excellence is through careful and deliberate design and execution of service management processes and systems to improve flow-through. That means increasing the speed, efficiency, and accuracy with which inputs are acted upon to achieve a desired result.

In the telecommunications business, a long history of monopolistic management permitted service providers to ignore the inefficiencies of their operations processes and to embark on wasteful "point solutions" that met individual departmental needs without regard to the impact on overall company performance. Reversing the trend is painful and extremely difficult, not to mention expensive. Stock watchers will note the trend of service providers to take significant write-downs against earnings in order to make the transition.

What is Service Management Excellence and Why Is It Important?

Communications is a service industry. In advanced economies, the service sector continues to rise in importance as a source of foreign exchange, jobs, and sustainable national competitive advantage. Companies in the service sector need to excel and provide leadership in the levels of service they offer to their customers if they are to survive. Many service industries have known this for a long time, and pioneers such as Federal Express, L.L. Bean, British Airways, and McDonald's are widely held as leaders in their fields. Viewing overall service to the customer as a source of competitive advantage is new to many in the communications industry, but those who aspire to be the best in this field will need to demonstrate excellence in meeting continuously rising customer expectations.

Over time, this won't be a new issue for communications. It will be in the bloodstream of the industry as it is in other successful service companies. But right now, achieving excellence in service management is a novel concept for many communications companies. Many do not yet even know that it needs to be done. They face a rude awakening against competitors who have learned that it can and must be done.

Many industries are littered with the casualties of major discontinuities—those undisputed market leaders who failed to emerge from a major industry shift still in pole position. The incumbent players see the need for change too late, and by the time they respond the race has been lost. When the radial tire displaced cross-ply, when color televisions replaced black-and-white, when silicon transistors displaced germanium, when distributed computing displaced mainframes—in these and hundreds of other examples, the established lead player failed to survive the discontinuity without being severely bruised. And the communications industry is going through the largest discontinuity that anyone could ever imagine. Those companies that fail to perceive that excellence in service management is crucial to their survival probably won't make it. There are no quick fixes, no magic formulas, no quality pills. In the words of L.L. Bean's president Leon Gorman, providing excellent customer service is "just a day-in, day-out, ongoing, never-ending, unremitting, persevering, compassionate type of activity."[1]

Exploiting process flow-through in achieving service management excellence

As we explained earlier, service providers go through a number of phases in facing up to the realities of a competitive marketplace. When restructuring and downsizing have run their course, the reengineering phase becomes crucial. At its heart lies a thorough examination of all of the company's core business processes from end to end, with the "ends" of those process chains

[1]Uttal, Bo. 1987. Companies that Serve You Best. *Fortune*, December 7:98.

often lying outside a service provider's boundaries. Customers might wish to raise an order for new service electronically from their own ordering systems and send the order to a service provider automatically. In many cases, the fulfillment of that order cannot take place without multiple service providers being involved. This is very typical in the United States, where the market is currently fragmented between local and long-distance suppliers, but it is also becoming a commonplace need within the global service provider alliances. Inside the service provider's own operations, many functions and subprocesses might be required to fulfill the order. Is the customer creditworthy? Do we have network capacity? Is this the first service of this type that a customer has or is it more of the same? Does the customer want separate or integrated billing?

Reengineering philosophy says that all of the steps of a process should be considered on an end-to-end basis. The aim is not small incremental improvements to an existing process but the large step changes in effectiveness that come with a radical "out-of-the-box" redesign. Champy and Hammer challenge the conventional wisdom dating from the industrial revolution that processes need to be broken down into a myriad of small steps, each involving a very simple task for the human beings who make up the process chain. They argue that in the world of advanced information systems, with complex and sophisticated applications, processes can be redesigned into a smaller number of steps with computers, not people, as the primary building blocks. They use the term *disruptive technology* to highlight the fact that new technologies can permit a radical rewriting of the rules by which companies can organize their processes. Table 2.1 gives a few examples of their ideas as they might relate to the communications services industry.

TABLE 2.1 Examples of Disruptive Technology and Its Impact on Process Design

Old rule	Disruptive technology	New rule
Information can only appear in one place at a time.	Distributed databases	Information can appear simultaneously in as many places as it is needed.
Business must choose between centralization and decentralization.	Distributed systems	Business can simultaneously reap the benefits of centralization and decentralization.
Field support people need offices where they can receive their work orders.	Work management systems with online access from notepad computers	Field personnel can be directed from wherever they are, thus improving productivity.
Network people look after the network, customer service people handle the customers.	Object-oriented representation and distributed computing environments	Data can be viewed by different people and systems for different purposes, allowing greater organizational flexibility.

What most organizations don't do very well is think through the impact of new technologies and how they can be used to work smarter and attain service management excellence. In our experience, telecommunications companies can be some of the worst at implementing new, advanced network services for their own use. It's surprising, for example, just how many service provider companies' employees have to get special dispensation to connect to the Internet or how frequently e-mail messages are bounced by a service provider company because their internal networks aren't properly configured. But then, Massachusetts Institute of Technology, one of the world's leading centers for advanced technologies, had rotary dial phones until a couple of years ago!

Wise and widespread use of technology is central to solving the "three-legged stool" problem. Unfortunately, its implementation is far from simple. Service providers can have tens, hundreds, or even thousands of discrete computer systems and applications that form their legacy base of systems. A logical, highly optimized process flow designed on paper can be a nightmare to turn into an implemented reality across such diverse systems. This is where new entrants into the telecom marketplace have a distinct advantage. They can design their processes in this highly automated process flow-through manner because they do not have legacy systems. They can achieve greater efficiency and therefore a lower cost base while at the same time offering higher levels of customer service and faster reaction times to the marketplace.

The Lean Service Provider

Mastering the skills of radical reengineering and, perhaps more importantly, growing the skills and competencies to implement the outcomes will be the hallmark of the lean service provider. The lean service provider, like its close cousin the lean manufacturer, is as different from the bureaucratic monopoly service provider as the "fly-by-wire" Boeing 777 is from a wood-and-canvas biplane. The lean service provider is one who has mastered the art of service management excellence, one who balances the three-legged stool continuously in the battle for market share and growth in increasingly competitive markets. The lean service provider is one who can continually make major operating improvements by process reengineering and can turn those reengineered, end-to-end process designs into practical reality—quickly. The lean service provider is one whose people can be unleashed from routine process implementation, caught up in an ever-increasing swirl of paper and procedures, to do the tasks that computers can never do—plan the next steps, innovate, and delight the customer.

The lean service provider delights not only the customer but its shareholders as well. In a world where incumbent service providers will inevitably face declining market share and significant falls in price, the lean service provider uses the talents of its people to grow the overall market and relies on automated, end-to-end processes to slash its operating cost base. The European car industry is now finding out what the U.S. auto industry learned a few years ago and Toyota and others learned in the 1950s: Radical change towards lean production is not an option—it is essential for survival. So, too, will the world's communications providers find that becoming a lean operator is a matter of corporate survival. For those companies that haven't yet faced competition, take note: It's far better to build the competencies now, before facing the competitive tidal wave. Those already experiencing the pull of the competitive current will need to remember that the number of competitors they face now is nothing like what it will be, and they should take steps now to build on whatever lead they may have. End-to-end service automation is easy to say, but very tough to do in a hurry.

How lean service providers approach service management automation

Much of the activity underway within service provider companies today involves automating and streamlining the interfaces between internal systems, since total process reengineering efforts have identified many opportunities for improvement. For example, a major European provider studied the process involved in providing a 2Mb private circuit, which took an average of 60 days to complete. They found that the amount of measurable "real work" took only 12 hours on average, with the rest of the time taken up by handoffs from one department to the next, one system to the next. Their reengineering objective was to reduce the total service provisioning interval to an absolute minimum while at the same time reducing the cost of the process by eliminating unnecessary manual overhead.

When evaluating internal processes and trying to improve process flow-through, tremendous improvement can be realized by applying well-accepted quality principles, such as reducing the number of steps in any one process, improving the quality of results of each step in a process, and focusing actions where the problems are most severe. But what are the unique challenges to this industry?

When Service Management Is Your Business

When what you sell is a managed communications service (one that is a complete package of capabilities, such as a virtual private network), your management capability is part of your service offering. What distinguishes the problems in managing communications services from those experienced by companies that provide other types of services, such as transportation or financial services? The complexities involved in communications service management far outstrip any other industry, for several reasons.

1. Telecommunications involves an extensive service delivery chain that is becoming more complex as the industry changes shape. The delivery of a single service often requires the service provider to interact with other providers to obtain geographic coverage or capacity and increasingly involves automated links with customers. And while it might be argued that these same characteristics can be found within global banks, service providers have an additional delivery-chain dependency on highly specialized communications equipment. While in many cases the provider might own and operate this equipment, most service delivery processes result in some change being implemented in real time on the networking equipment, demanding a close working relationship with equipment suppliers to make those changes. Sadly, such close cooperation is rare, with many service providers relying on traditional, formal, arm's-length procurement arrangements.
2. Service providers operate on the leading edge of new technology, while dragging along plant and processes that can date back 100 years. Regulatory requirements or government fiscal policies often encouraged an investment strategy and depreciation schedules geared to a much slower-paced industry than the one in which service providers now find themselves. Making the transition to a lean service provider state, while having to accommodate equipment and operational support systems that are still "on the books," adds extra challenge.

   It is common for service providers to create "overlay" networks and processes in order to bring new services online, even as the company struggles to remove existing process headaches. For example, in order to introduce a new service within 12 months of product definition, it might be necessary to install a more advanced switching and signaling capability, as well as all-new ordering and billing systems that operate "on top of" the main systems and are used exclusively for that service. These overlays can violate every possible quality principle, and they inhibit the move toward becoming a lean service provider, but they are often necessary to stay competitive. Some service providers faced with this reality are basing their overall process design around these new service overlays, with the thought that existing "legacy" services can be migrated to the new processes and systems more easily once there is a core of people familiar with operating under the new rules.
3. The communications industry is becoming increasingly distributed as geographical boundaries are stretched. The likelihood that people involved in one step of a process will be collocated with the people operating the next step is very low. Making a change in a process in which the cooperation of multiple groups or systems is required is extremely difficult, both politically and technically.

In addition to these points, service providers face the same problems as any other service company when trying to reengineer their business. For example, companies often apply their financial controls and budgeting process in the wrong places, starving the internal process-reengineering activities of cash just when they need to expand to move the company toward the lean provider state. As Louis Gerstner, now chairman of IBM, remarked when he was president of American Express, "Because of the structure of most companies, the guy who puts in the service operation and bears the expense doesn't get the benefit. It'll show up in marketing, even in new product development. But the benefit never shows up on his own P&L statement."[2]

Too true, unfortunately, and companies that aspire to service management excellence need to understand this point fully. During the early phases of competitive evolution, when an operator squeezes cost out of the business, all too often it can squeeze the resources required to take it to the lean stage. Only enlightened managers who can see the whole picture know where to cut and where to invest.

Unfortunately, most managers can't see this overall picture and are unable to make decisions to invest in order to save. Even companies with a fairly well-articulated overall vision of how they need to transform their companies' processes are often stopped dead in their tracks by the fairly high initial expenditure required to implement the vision. This hesitation is particularly evident in how service providers are responding to their suppliers. Instead of demanding adherence to critical standards and making the initial investment of time and money needed to back up their demands, service providers usually allow short-term financial worries and new product pressures win the day, thus postponing the time when real improvements can begin and increasing the base of nonstandard equipment and systems that will eventually need to be modified to meet the new requirements.

In the main, process reengineering efforts are considered highly proprietary, and decisions about how to construct business processes are kept confidential. Service providers are in general agreement that if they get their internal processes "right," they stand a very good chance of competing successfully for business in the coming years. For this reason, service providers have no interest in reaching industry agreements on how their processes should be constructed or how their internal ordering, billing, and operations support systems should be built.

However, in two key areas, service providers face barriers that they can't surmount by themselves. One of these has to do with the lack of a sufficiently scaleable computing infrastructure to support the types of processes they de-

---

[2]Making Service a Potent Marketing Tool 1984. *Business Week*, June 11:170.

sire. The other involves the fact that nearly every internal process also involves points of contact with outside entities, including customers, other service providers, and suppliers. Without a certain level of industry agreement in these two areas, service providers are prevented from achieving maximum process flow-through or applying technology to automate key service management processes. Later in this book, these industry needs will be explored more fully.

Chapter

3

What Is Service Management?

So far, we have explained that achieving service management excellence is essential and hard, but we haven't explained what it actually is. Put two rather generic words together, like service and management, and they can mean almost anything, depending on who is using the words.

The term *service* can apply to serving a meal in a restaurant, fixing a car, or providing financial transaction services. In this book, the service is applied to communications service, such as would be used by customers to exchange information, conversations, or pictures over a distance.

The term *management* is an equally generic term. It can be applied when describing the art of making a profit (business management), the skill of keeping a family clean and well-fed (home management), or the ability to keep the streets free of garbage (sanitation management). In this book, management refers to the set of processes and activities necessary to deliver communications or information services to customers and operate them in a way that meets quality and cost objectives.

Service Management Basics

It might be easier to understand service management if it is first viewed in relation to other management functions within a service provider organization. The premier global standardization body for telecommunications, the International Telecommunications Union (ITU), has agreed upon a common model to depict management in a communications environment. The main framework is based on work originally undertaken by a BT team led by Keith

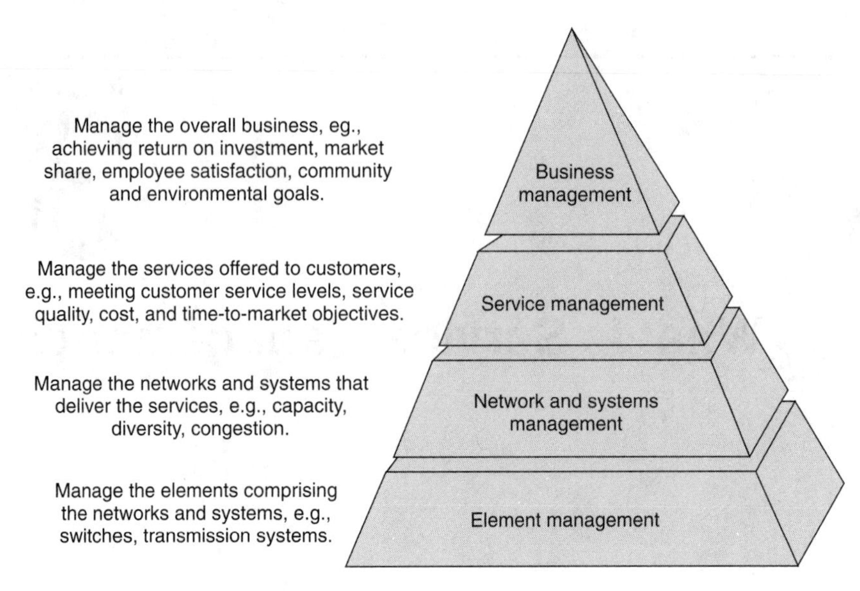

Figure 3.1 The four layers of management.

Willetts in 1987 and is now widely accepted.[1] Known as the Telecommunications Management Network (TMN) framework, it shows four layers of management, as shown in Figure 3.1.

This model is simply a way of breaking down the complexity of a service operator's world into more understandable chunks. It is based on a simple value chain, with the break points being typical trading or organizational boundaries. Like all architectures, it is not intended to be a precise science, but rather a framework for discussion. It is, however, proving to be remarkably robust and adapts well to the emerging layering and niche targeting of an industry that was once vertically integrated. Service providers are emerging that don't own or operate networks. Carriers' carriers are emerging that deal only in basic transport networks. And in some countries, regulators are forcing either legal or financial separation between these layers in order to open up markets to competition.

We have found that the layering in this model is sometimes misunderstood, so it's worth taking a closer look at each layer in turn and being clear what we mean by each one.

Most people use the model bottom-up, but its real value is to start top-down. Only by looking at the overarching business management issues—such as we examined in the first chapters—will the need for investment in

[1]British Telecommunications plc. 1988. Open Network Architecture—Communications Management Architecture Release 1.0. DS0009, Part 1.1.

the lower layers become clear. It's also the only way to see the requirements placed on each layer by the one above it.

As shown in Figure 3.1, the first management task relates to managing a communications business—the financial imperatives, meeting the needs of shareholders, customers, employees, and society. From these goals and objectives comes a framework for managing the services that the company delivers to its customers—the entire customer-service spectrum, including order handling, service quality, problem handling, billing, service development, and so on.

These services are derived from a network infrastructure—either owned and operated by the service provider, if they are vertically integrated, or provided by another organization. This network management layer is responsible for ensuring end-to-end connectivity, network integrity, capacity planning, disaster recovery, etc. In addition, the network management layer needs to ensure that any failures that might be service-affecting are immediately reported to the service management layer, as part of meeting service level agreements.

Finally, the network itself comprises many elements—switches, transmission systems, etc.—from many suppliers. These need to be managed for optimal performance—error rates, alarms, performance characteristics, and so on. Failures at this level can have an impact on the network level, which in turn can have an impact on service to the customer.

The most important aspect of this TMN model, then, is that the business performance of a service provider is dependent on the excellence of its service management. In turn, service management relies on the excellence of network-level management, which in turn relies on the underlying element management. Hence a pyramid, with each layer depending on the layer below. So, although this book focuses on the business drivers and service management aspects of a provider, it is not meant to imply that network management and element management are not vital—they are.

Our purpose in examining this model is to highlight for those lower down this value chain—for that is what the TMN model is—that decisions made by a supplier of communications technology with associated element management capabilities can directly impact levels of service and business performance. In our experience, many suppliers cannot understand why the management aspects of their technology are so important to a provider and why they need to be integrated into a holistic management scheme. The answer is simple: dollars, pounds, yen, francs, deutsche marks! Unless this model is understood and correctly translated into systems, processes, and procurements, then the provider's three-legged stool will fall over. The more astute suppliers are beginning to understand this, while the less astute may find themselves explaining it to their liquidators!

A Detailed Look at the Management Layers

Distinctions are made between these management layers based on the functions performed. Put another way, it is possible to distinguish between business, service, network, and element management by the types of questions about which managers at each level might be concerned.

Business management

One of the overarching goals of business management (Figure 3.2) is to improve profitability and hence the return for the shareholder.

Questions asked at the business management level might include the following:

- How can we grow earnings-per-share when faced with declining prices and market share?
- What businesses and markets will we focus on?
- With whom should we enter into alliance or partnership and for what business purpose?
- How do our costs per unit (number of employees, revenue, or number of customers) compare with our competition?
- How can we allocate true technology costs against the business sectors to identify the real costs of our business services (as is increasingly being demanded by regulators around the world)?

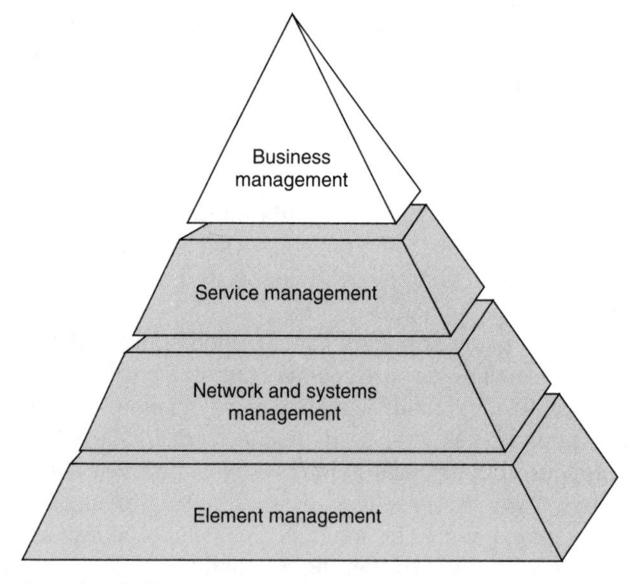

Figure 3.2 Business management.

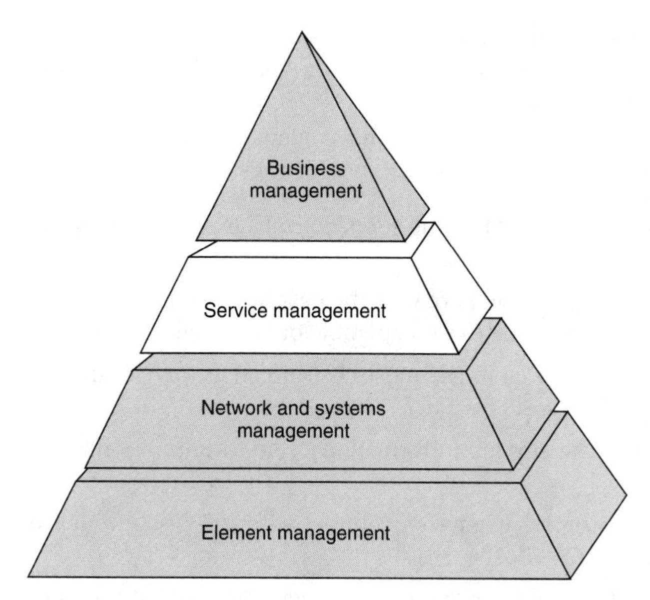

Figure 3.3 Service management.

- How can we stimulate an increase in revenues?
- What new products (service offerings) warrant our investment, and how can we reduce their time to market?
- How will we respond to competitive threats?

The kinds of information that are important at the business level are those that can be looked at across product lines to determine which products (service offerings) are successful, which sectors are doing well, where expenses are on the rise, and how overall performance compares to the competition. At the business level, the systems tools are likely to be fairly generic in nature, involving accounting, financial, and product-tracking applications that could be easily applied to any other industry. Understanding the types of data needed at this level and the form in which it is desired is crucial to the service management level, which generates much of the information used in business analysis.

Service management

In the service management area (Figure 3.3), the key goals are to improve customer service, reduce costs, and shorten time to market. In other words, it falls to service managers to find ways of achieving overall profit improvement through better operational processes and the creative application of networking and information technology. By service management, we mean all of the processes and systems employed to deliver services to customers and manage them through the service life-cycle: service creation, order handling, customer administration, marketing, problem handling, billing, and so on.

To test where they are in achieving service management excellence, questions asked by service managers might include these:

- Are we consistently meeting or beating service level agreements with our customers?
- How do our service offerings compare with the best in the industry? And if they don't, what must we do to offer comparable or better service quality?
- Are we operating our services in the most cost-effective way? Could we reduce service operation costs further?
- How can we shorten the service introduction cycle to improve time to market?
- Do we delight our customers when they interact with us? Or do we leave them reaching for a competitor?
- Are we handling customer requests promptly, accurately, and at the minimum possible cost?
- Can we guarantee end-to-end quality of service at target costs even where we are reliant on other providers?
- Are we aware of problems before our customers tell us? Can we find and correct those problems in the fastest possible time?
- Can we extract information about our customers from the entire range of service management systems for marketing advantage or service differentiation?

At the service management layer, the objective is to make the link between network performance and the service levels offered to individual customers. Thus, the systems found at this layer are a mixture of control systems for such things as network manipulation and service performance monitoring and more typical business systems such as ordering, billing, and problem-handling systems. This mix of network-facing and customer-facing processes and systems is examined more closely in the next chapter.

Network and systems management

Although network management and systems management often involve different organizations (Figure 3.4), both play a full part in delivering improved profitability by providing service management with an excellent service infrastructure. Increasingly, services are provided through a combination of traditional networking technology and computing systems. For example, if a service provider gives its customers access to a system

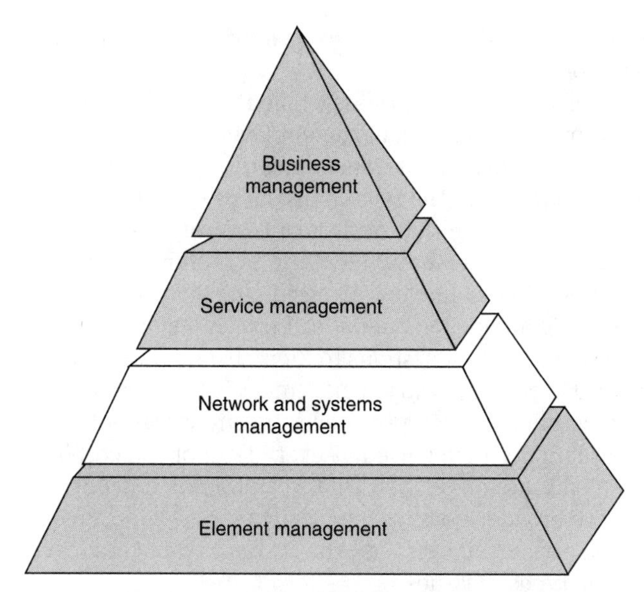

Figure 3.4 Network and systems management.

through which they can modify their service configuration, maintaining the health of that system is every bit as important to delivering the service as maintaining the network switches and facilities.

Network managers are concerned about ensuring connectivity and performance objectives across the network elements. They might well ask:

- Do we have the network configured so that traffic loads can be handled as effectively and efficiently as possible?
- Do we have a complete, up-to-date inventory of our networks and systems, and are we using these resources most efficiently?
- Do we have a disaster recovery plan in the event that we lose part of the network?
- Is the network able to adjust flexibly to changes in demand?
- Have we made all the necessary routing changes to implement a new calling code?
- How should we alleviate congestion—by blocking call attempts, rerouting, or adding capacity?

At the network management level, systems are typically on the receiving end of detailed data about the health of individual elements and the ability of individual switches to complete calls over their many trunk routes. Using this data, together with a complete network topology, the network management systems determine where significant blockages exist and either take action to reroute traffic or initiate a work order to correct a problem. The output of these systems is a higher-level (less-detailed) summary of information that can be used by the service management layer.

In managing a computing infrastructure, including the systems that are used to deliver service management, the related discipline of systems management has the objective to ensure that performance across the systems elements meets service level agreements. Increasingly, service providers are using commercial computing technology alongside the more specialized telecom switching fabric to handle elements of call processing and to provide easy-to-use service management systems to their customers and operations centers. This small slice of systems management is but a fraction of the job faced by those within a service provider company whose job is to manage the many information systems used at every level of the corporation. Later, as part of the discussion on systems-level integration needs, we touch upon some of these broader systems management issues.

As part of supporting service management, the systems manager is concerned with the following types of issues:

- Is the overall response time to the user of a specific application meeting or exceeding the service level agreement?
- Are our backup and recovery procedures adequate?
- Which applications belong on a distributed, mid-tier computing environment, and which are better suited to the mainframe?
- What can we do to handle user churn more effectively?
- What's the transaction rate threshold for the overall system?

Element management

Managing the elements takes place both in the systems world and the network world, and the lines are increasingly becoming blurred as networking equipment incorporates more and more computing capabilities and as computers are used increasingly to perform traditional network processing functions. Regardless, element management functions (Figure 3.5) are responsible to see that individual devices are properly maintained and to take action in the event of failure.

Element managers might ask:

- Have the routing tables been updated?
- How can we relieve system access congestion?
- Why is the switch blocking calls to that area?
- What response time does the application deliver, and how might it be redesigned to gain improvement?

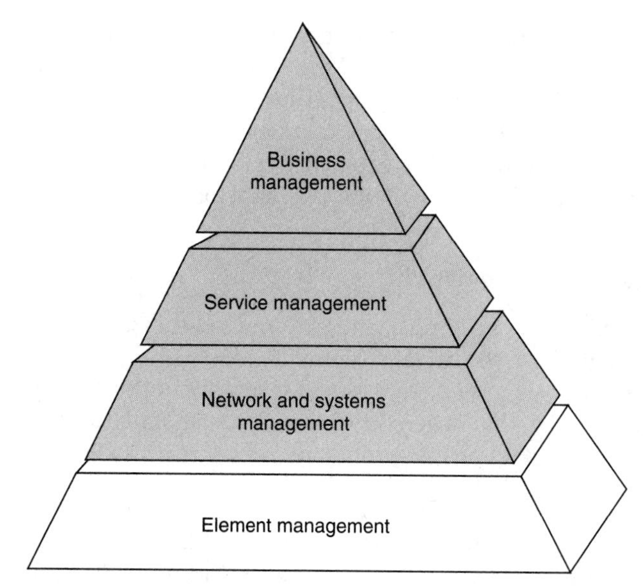

Figure 3.5 Element management.

- Has the transaction rate threshold been exceeded?
- What is causing the signal processor to fail, and did the standby unit activate automatically?

Until recently, many equipment manufacturers provided very little support for managing their equipment elements. Service providers built their own systems to fill the gap, often not even realizing the levels of investment they were making in order to gain the level of control and visibility they required. About a decade ago, following the first waves of deregulation and liberalization, the managers of private enterprise networks began to buy and manage their own networking equipment, seeing the opportunity to innovate faster than the providers and reduce their costs. What these private enterprise network managers hadn't anticipated was a lack of element management tools from equipment providers, but, unlike the service providers, they were unwilling or unable to build their own element management systems.

The demands of these private enterprise network managers, aimed initially at the LAN equipment vendors, made an impact on the suppliers of large-scale networking equipment as well. Within the last decade, most equipment suppliers have come to grips with the need to make their equipment more manageable and have provided applications or full systems to do the job. However, their priority has been on making it possible for the technician to monitor and repair the unit—not on the need for information to be passed to the next higher layer in the management chain.

Still, today, we see equipment vendors discussing the management of ATM or SDH networks, seemingly unaware of the fact that there is more than one use of the outputs of their management information. Impressing upon the equipment suppliers the importance of an element management capability that can serve the wider needs of the network and service management systems continues to be a difficult but essential task.

Why Focus on Service Management?

If all the layers of management are important (and indeed they are) why does service management deserve special attention?

There are several reasons why service providers worry about the service management layer more today than when the industry was under less pressure to deliver customer satisfaction and profitability:

1. As Laura Cerchio, an executive responsible for software technologies in the area of network and service management for STET/CSELT, explained, "Network management is something we have always had to do. Service management is new to us, and it's urgent that we learn what's needed and how to do it if we are going to stay in business after we are liberalized." (CSELT is the research organization of STET, which is the holding company of Telecom Italia.) Service providers have years of experience performing network and element management tasks reasonably well but only limited experience in managing from a service perspective in a competitive environment that increasingly calls for excellence in this area. Many people involved in reshaping service provider companies readily admit that "they don't even know how much they don't know" about what it means to achieve service management excellence, and how to get there.
2. With increased competition, service providers are acting more like other businesses, and the need for service management excellence—to continuously improve service capability while reducing cost and time to market—has begun to gain importance and visibility. Managing to service level agreements highlights the need for a different way of measuring performance and new skills within the employee base.
3. Value-added service providers are rapidly springing up in many markets. These new entrants to the communications service industry are proving that it is possible to make money without operating a network, using the services of traditional operators on a "carriers' carrier" basis. Service delivery chains are becoming more complex, and their business implications are causing service management to take center stage.

Public Service Providers: The Big Investors

When looking at service management, there are differences between the needs of companies that provide communications services as their primary business and those that use those services as a business tool. Many managers of large private networks have employed the concept of service management for a long time, and they have been lobbying for many years for improvements that will let them manage services to their end users with less manual intervention and better quality. But service management among public service providers will be a battle for survival and will account for significantly more investment than the enterprise networks. So the real pressure for changes to manage at the service level—backed by hard cash—is coming from the public service providers.

A global airline, for example, views network and computing technology as one of several crucial resources to be used to meet business goals and holds a senior officer responsible to ensure that company-wide communication and data exchange happens efficiently and without incident. Other officers are responsible for delivering transportation services to the airline's customers, and these activities are center stage in the airline's boardroom.

A public service provider, on the other hand, makes money by delivering network and information services. Instead of being considered just one part of a larger business equation, service management, as we have defined it in this book, is their business. Product investment decisions made by service providers are directly reliant on the company's ability to manage service development and delivery well. Contrast this with an airline, whose dependency on communications and information services (although crucial) is but one part of the equation in providing high-quality transportation services to its customers.

Why is this distinction important? To companies in the business of selling communications equipment, management platforms, or management systems, knowing where the money will be spent, and for what, is key to growth. The public service providers comprise some of the largest companies in the world, and taken as an industry, they account for billions of dollars of investment in telecom equipment, computing equipment, and software each year. Looking only at the developing countries, something on the order of $200 billion has been allocated for telecom equipment over the next five years.[2] A conservative estimate puts computing expenditures by service providers at more than $50 billion annually. Those involved in setting standards for manageability of equipment would do well to listen to this market segment especially closely.

---

[2]Arnst, Catherine. 1995. The Last Frontier. *Business Week*, September 18:99.

The positioning of service management—between business management and network management—illustrates quite effectively the role it plays within an organization whose main mission is the delivery of network-based services. In companies that make their profit from selling communications services, the discipline known as service management is where the firm's business objectives (profit, market share, or customer satisfaction goals) are applied to the use of the organization's assets (the network infrastructure).

Chapter

4

Looking Inside the Service Management Layer

The two major parts of service management are shown in Figure 4.1. One part, called "customer care," deals with the many processes needed to actually deliver services to customers, such as order handling, problem handling, performance reporting, and billing/collections. The other part, called "service development and operations" deals with the manipulation of the underlying network asset to create new capabilities and to monitor performance at the service level. Both must be superbly managed, and they must be linked if the service provider is to thrive in the competitive hothouse.

What are the key management issues in each of these parts of service management? And more specifically, what are the implications of those issues on the development of service management systems?

Customer Care Management

Providing good customer care points to the following needs:

- Providing current and accurate information to customers.
- Delivering services when promised that are right the first time.
- Resolving problems quickly and keeping the customer informed of status.
- Meeting stated service level agreements for performance and availability.
- Providing an accurate, easy-to-read bill in a format and currency that the customer wants.

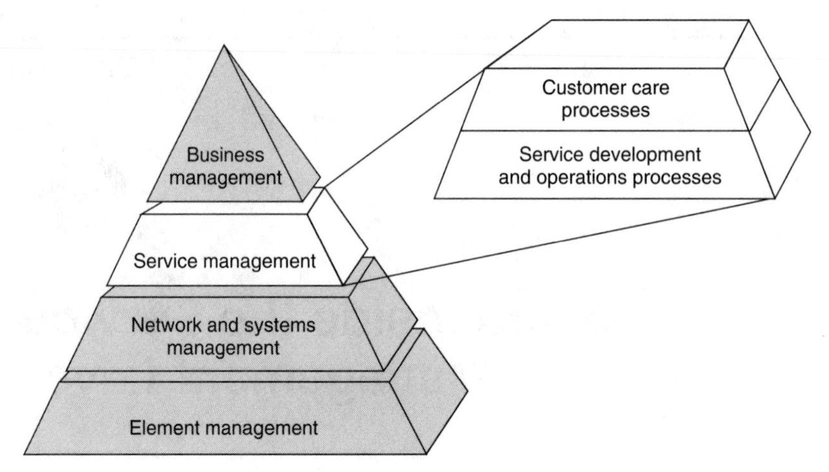

Figure 4.1 The two parts of the service management layer.

- Reflecting billing adjustments automatically in the event that service performance did not meet objectives.
- Providing relevant performance data in a format the customer can use.
- Ensuring that the customer experience with the provider is always delightful.

There are a number of reasons why customer care processes benefit from end-to-end automation. In the following examples, we examine the customer care side of service management, as shown in Figure 4.2, and why it is so important.

Improved customer image

Automation of customer care processes enables employees to perform their jobs quickly and professionally. When it is possible to enter an account number or name and see not only the latest bill, but an accurate, up-to-date record of all pending orders, current reported problems, and usage history, the service provider's contact center can be responsive, helpful, and quick in answering questions. Instead of being trained in the art of "knowing who to call" for information, contact center personnel can focus on the manner in which they treat the customer and the opportunities to sell additional services.

Contrast the following two scenarios:

1. When a customer reports a problem to service provider A, the technician requests lots of information from the customer to identify who is calling and from what site. Following what seems an interminable period of pre-

liminaries, the technician leads the customer through a series of questions about the problem in an effort to pinpoint the cause. Eventually, the customer says "Could this have anything to do with the installation work you just finished yesterday?" The technician replies "What installation work?" The customer spends another lifetime explaining what has been done and what the service configuration looks like (fortunately this customer is knowledgeable enough to do so!) and finally hangs up in disgust, not at all confident that the technician could manage his way out of a paper bag, much less put two and two together to come up with the answer. And behind the scenes, because the technician has made a mistake entering the customer information, there is no record of a problem report associated with the customer's service level agreement (SLA)—another source of irritation later in the month!

2. When this same customer reports a similar problem to service provider B, the technician asks for the name of the customer and immediately accesses the customer's record, which is complete and up-to-date. The technician says "I see that some work was just completed there which might be causing this problem." The customer is impressed. What's more, the technician initiates action to resolve the problem immediately or commits to a due date and launches a work order, not only delighting the customer but avoiding any rebates due.

It has been proven that delighting customers is the key to customer retention. Any customer that is not delighted is at least open to the competition if not actively shopping.

Exploitation of customer data

There is another advantage to automating customer care processes, which is the ability to create new service offerings based solely on the manipulation of customer data. An obvious example is the creative use of the billing system. With no new network capabilities, it is possible to deliver new service options simply by applying discounts to customers' bills based on usage. A classic example of exploiting the billing capability is MCI's Friends and FamilySM service, which links the billing records of multiple customers

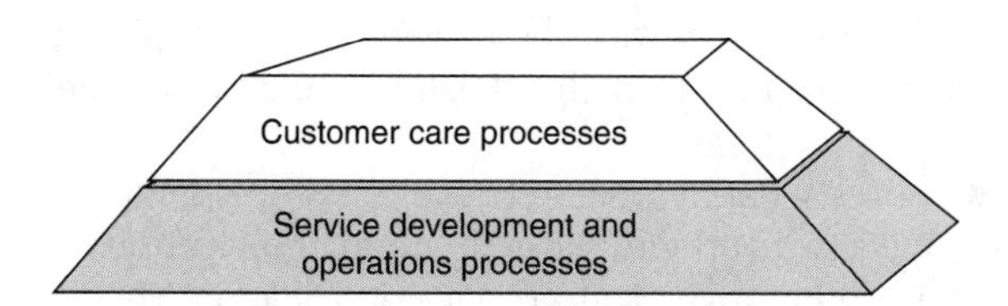

Figure 4.2 The customer-facing part of service management.

as part of a calling group and provides discounts based on calls placed among those specific numbers.

On-demand information

Improving customer care process linkage can easily be extended to the customers' systems, and the benefits of automating these links can be considerable. For one thing, providing automated access to service information means that the customer can get the information he or she needs, when it is needed. A good example is problem-status information. Today, in an effort to be customer-driven and proactive, many service providers have begun notifying their customers of the status of all reported problems at preset intervals. This can have one of two effects on the customer, but both can result in annoyance. In the first instance, the customer may want information more frequently than it is given and so may be frustrated by having to wait for periodic updates. This customer will probably still spend a lot of his or her time (and the service provider's time) calling to find out what is going on. In the second instance, the customer may not be interested in knowing the status—the problem might not be critical in his or her view—so the repeated calls are annoying because they take time and deliver no perceived value. By giving the customer direct, on-demand access to the information, the customer can control when data is requested and can get exactly what is needed. The service provider personnel will spend less time answering the phones or initiating calls, and everyone will be happier.

In a service trial a couple of years ago, some startling and surprising results were obtained. The customer, a global bank, leased a large number of 64K and 2Mb private circuits from two major service providers. One of the providers offered, on a trial basis, full online access to service management information, giving the customer on-demand real-time status of the links, online fault reporting, and progression status. Now that the customer could see if the links were okay, they didn't blame the service provider so frequently when their internal banking services went wrong. In fact, the customer's perception of service quality was that there had been at least a 50% improvement in link quality when actual quality had remained constant. Disputes over out-of-service statistics fell by 95% because both the customer and the service provider had a time-stamped audit trail of outages and restorations. The customer was so delighted by the improvement in customer care that it placed more orders with the provider that was giving advanced levels of service management, to the detriment of the second provider, which wasn't.

The service provider was pretty happy, too, because it saw a significant reduction in the number of calls to its operations center and equivalent cost reduction from avoiding the need to investigate reported problems. This is a good example of the 1:10:100 rule used in manufacturing or systems design. If a problem can be fixed during the design phase (in this case providing information and therefore avoiding a complaint), it costs 1 unit of expense. Fixing a customer's problem online costs 10 units, while sending a technician to investigate represents 100 units of cost. So, for every call that can be resolved using an online capability, the service provider saves money while producing a happier customer.

Service Development and Operations

The other aspect of service management, service development and operations, involves making creative and economic use of network assets, presenting different issues that are just as important as customer care (Figure 4.3).

Doing a good job of service development and operations means:

- Implementing a cost-effective service design that can be brought to market quickly.
- Building in the ability to configure, install, and bill for the service within stated goals.
- Monitoring service utilization to prevent service impairment.
- Providing the ability to tie network performance to individual service level agreements.
- Managing the quality of a specific service as delivered to all users of that service.
- Imposing sales restrictions if and when customer demand exceeds the ability to meet target service performance levels.

Automation of the service development and operations processes involves creating linkages with the many network and element management systems that operate at the next-lower layer of management within the TMN model. Let's look at a number of examples of these linkages.

Linking network performance, service performance, and customer experiences

A major U.S. service provider recounted the following horror story as an example of how hard it can be to link the "network" with the "customer." On

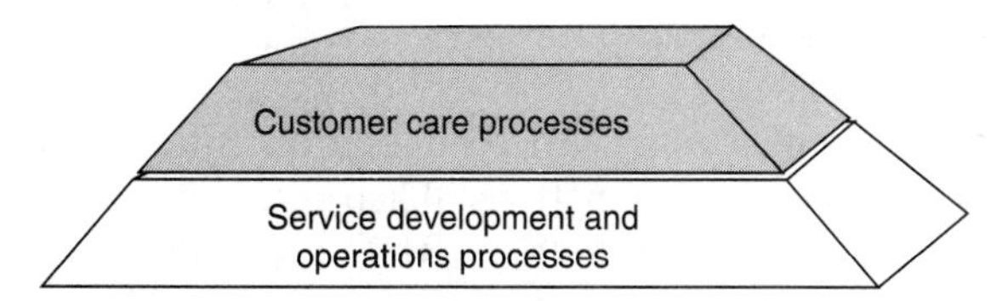

Figure 4.3 The network-facing part of service management.

Christmas Eve day (the busiest shopping day of the year), the service manager received a call from a particularly large and important financial-services customer. This customer made use of toll-free inbound services to link point-of-sale terminals with its fraud-detection unit. Although most credit checks were done automatically, some not fitting the profile of the card owner were bounced to operators who would verify that the card was, indeed, being used by the rightful owner and had not been stolen. Anticipating heavier-than-normal loads, the company had hired extra operators to be ready for this important business day. Suddenly, the center became very quiet as calls ceased. For 15 minutes or so, call volumes were reduced to less than 25% of the customer's capacity. At point-of-sale terminals around the country, customers experienced delays, and a significant number of them pulled out different credit cards in order to complete their purchase, costing the financial-services company millions of dollars in lost revenue. What had happened?

As a contrite service manager later explained to the irate customer, the person watching the performance of the service control point (not the regular technician, who was off for the holiday) saw that normal capacity thresholds were being exceeded as the database was queried for routing information to terminate calls to the customer's number. Although arrangements had been made to terminate additional traffic, the link between the service capacity increase and the network element (the service control point) was not made and regular default values were in effect. So, when the technician saw what looked like a problem, he blocked (some might say overblocked) calls to that destination number, with predictable results.

This could have been averted if the systems had been automatically linked. What was worse, because there was no link, the service manager was unaware of the problem until the customer reported it and so was unable to correct the problem as quickly as might have been done with immediate notification.

Achieving more for less

Linkages between the different service development and operations processes and between the service management and network management levels are needed if the service provider is to have any hope of achieving service management excellence. End-to-end process automation has the double advantage of improving the accuracy and speed of a task while also freeing personnel from routine jobs.

Let's look at a simple example of how a service configuration function might be automated. Upon receiving an order to establish residential service with three-way calling and call-waiting features, the ordering system passes account information to the billing system and sends configuration information to the service configuration system. The configuration system interprets what has been ordered, makes assignments of line equipment and a telephone number from available inventory, directs a software-controlled cross-connect to be made to connect the line equipment to the cable termination, and implements a software change in the switch to record the customers class of service and line features. At the same time, the billing system establishes the customer's account, performs credit checks, and so on. Both the configuration system and the billing system report the completion of their work back to the ordering system, which triggers personnel to notify the customer.

Not so long ago (and continuing today in many cases), this same simple function was performed by many islands of people, each using separate systems or manual records to advance an order from the order entry center to the line and number assignment center, to the plant assignment center, to the frame, and then to the switch. On average, as many as 10% of all simple orders encountered at least one error along the way, and at least five people "touched" the order before configuration was completed (not to mention the army involved within the billing group). A "good" service-delivery interval in the manual environment was a week or more for simple residential service. In an automated environment, configuration can be performed within minutes.

This example of service configuration process flow-through can be extended further to originate in a customer's system. Although once unheard of, service providers are providing their customers more and more control over service performance. Competitive benchmarks are rising fast, and all service providers should expect considerable pressure to open their networks to customers, at least to some degree, if they haven't already done so. Some service providers have been offering customer-controlled services for a decade or more. The advantages of automating end-to-end are in cost reduction as well as improved customer service.

However, making the link between the customer and the network management systems requires careful thought about terminology. Customers think in terms of the services they use rather than the language of the network device that delivers that service. For example, a customer could easily request that bandwidth be added between points A and B at a specific time in order to support a video-conference session or to facilitate the transmission of large amounts of data. The customer will not be able (nor should he be required) to specify which devices or facilities are to be modified, or how, but only needs to describe the desired end result. Creating translations between customer language and network commands is the obvious first step in automation, but service providers will gain accuracy and productivity if they use common terms as much as possible from the customer straight through to the networking equipment.

Terminology differences that exist today between management systems at the service, network, and element levels are generally there for historical reasons rather than by design, and process flow-through improvements can be realized if these are eliminated over time. Consider the plight of a service provider who makes use of switches from multiple vendors, each of which produces usage records in a different format using different terms. Although it is certainly possible to translate (service providers have been doing it for years), it adds unneeded complexity, which equates to time lost and errors made, not to mention unnecessary development cost.

Chapter

5

Service Management Requirements of Service Providers

So far, we've talked about the benefits of process automation in a general sense. But as service providers embark on service management improvements, they need to consider that there are two sides to every decision they make.

On the one hand, they are at liberty to construct their internal processes in any way that will place them at best advantage. The way they construct, simplify, and link their processes will either position them well for the future or will hold them back from achieving the kind of efficiency levels and flexibility needed to compete. On the other hand, they are not islands, but rather part of a service delivery chain that operates often in real time. The problem of receiving multiple usage records mentioned at the end of the last chapter can't be solved except through the cooperation of the switch providers. Links with customers can't be achieved except through interface agreements. And unlike many industries, service providers are often dependent on partners (which are likely also competitors) to provide end-to-end service, a trend that is growing with the emergence of alliance groups that are striving to provide sophisticated, seamless global services across a variety of infrastructures. Process linkage, then, needs to be viewed from both within the corporation and in terms of its links to the outside world. And, because processes are very interdependent, service providers are advised to consider both aspects as they develop their plans.

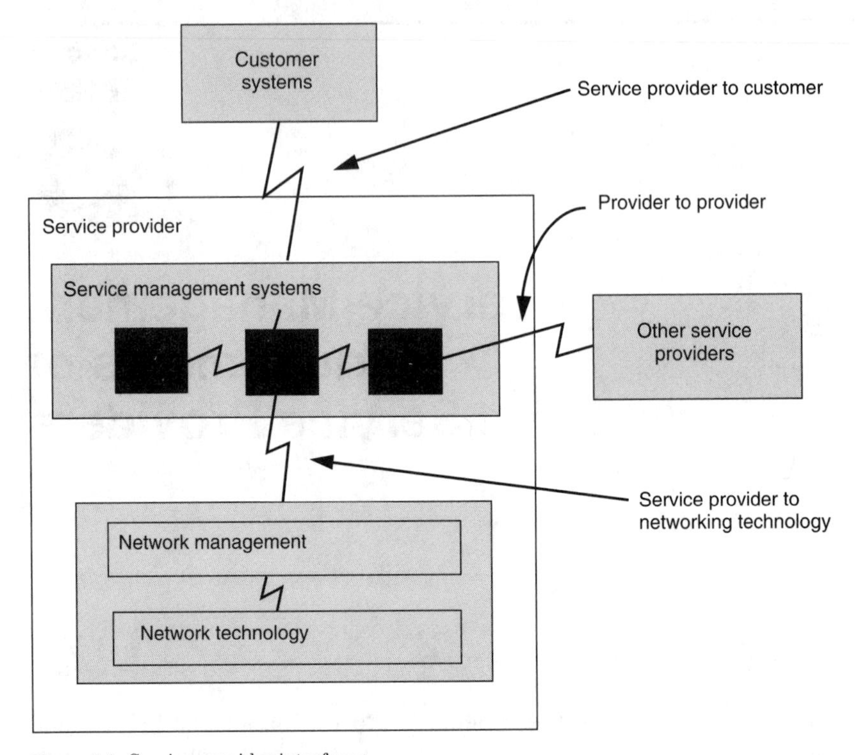

Figure 5.1 Service provider interfaces.

The Ins and Outs of Process Flow-Through

Only when a specific business benefit can be clearly identified and quantified will service providers willingly enter into common interface agreements. The types of benefits gained depend on the nature of the interface.

Consider Figure 5.1, which shows the points of interface from a service provider to its customers, other providers, and the supplier-provided network and systems technology used to create service capabilities.

The business objectives driving intercompany links vary. Looking at each of these interface points in turn, we can see the different business motivations that are likely to drive common agreements. We look at each in turn.

Process Flow-Through to and from Customers

Between the service provider and the customer, there are two key reasons for considering an industry-standard interface, as shown in Figure 5.2. The first is to improve the availability of information—its timeliness as well as accessibility. The second is to promote the use of more relevant information in a common form that makes it more useful.

Timeliness of information and ease of access

When they are having a service problem, customers want current status information regarding the provider's progress toward fixing that problem. Many service providers have begun to offer online access to that information via a terminal that connects directly to a system developed and controlled by the provider. While this is a good first step, customers with multiple providers can find themselves dealing with a proliferation of terminals.

Most large customers have their own help desk systems, which they use to log and track complaints from end users. Many would prefer to use these internal systems to access the provider's information. By actually linking the customer's and provider's systems—by extending process flow-through agreements to achieve full, end-to-end process automation—additional savings accrue, and customer satisfaction levels increase further by making availability of information easier and more timely for the customer.

Making information more relevant and easier to use

Another objective of reaching industry agreements with customers is improving the value of the information that's provided to the customer. Providing averages such as errored seconds, mean time between failures, and so on for the service as a whole is probably meaningless. The customer wants to know "How is my instance of the service performing against my service level agreement?" What's more, because customers tend to use performance information to compare multiple providers, they would like the providers to use common terms (with common meanings) in their billing records or performance summaries. Although some providers see it as risky to make it easier for customers to make direct comparisons, they will see

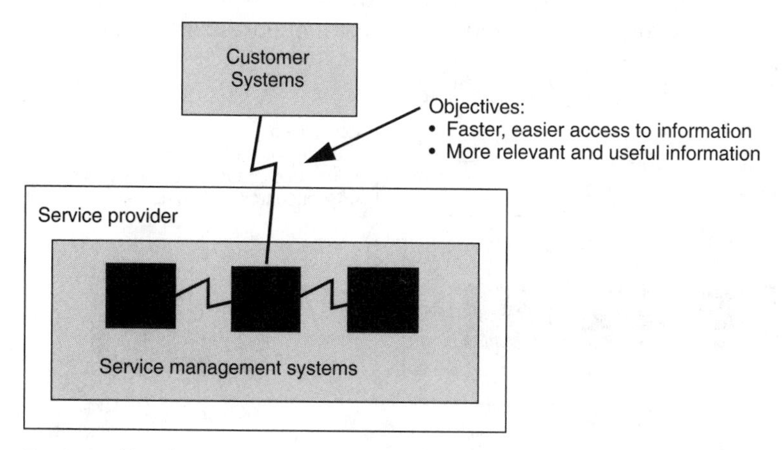

Figure 5.2 Benefits of automated process flow-through between providers and customers.

the benefit of using common terms firsthand as they increasingly construct services in partnership with other providers.

Making it easier for the customer to use the information that's provided is also important to most customers. They will want a wide variety of formats and layouts of such information and will probably want to further manipulate the data in some unique way, to feed their own network planning or expense-tracking systems, for example. Thus, presenting the information in a standardized form that can be accessed by readily available customer tools is a significant plus.

A key area of service differentiation for providers is if they can capture and package information that provides real customer benefit. There is a fine balance to be struck in differentiating services by providing commonly formatted information. Providers should seek to compete on the quality, accuracy, and benefit of the information they pass to the customer—not in its format or packaging. Customers may get service from more than one supplier, and they do not want to wade through widely different service information that cannot easily be reconciled and combined.

Provider-to-Provider Process Flow-Through

Increasingly, a single provider is unable to satisfy customer requirements using only its own facilities. Whether because of capacity, geographical, or regulatory constraints, service providers often need to contract with one or more other providers to supply capacity, features, or access. The subcontractors may very likely be competing for the same end customer, particularly as regulatory boundaries are removed and many providers are simultaneously eyeing the same global market.

In this competitive environment, what would cause service providers to agree to an automated interface? At least three compelling business reasons overshadow the natural reaction to make life more difficult, not less, for the competition, as summarized in Figure 5.3.

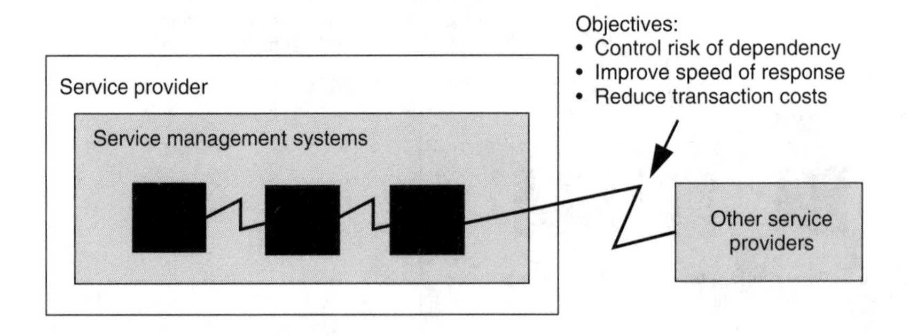

Figure 5.3 Business drivers of provider-to-provider process flow-through.

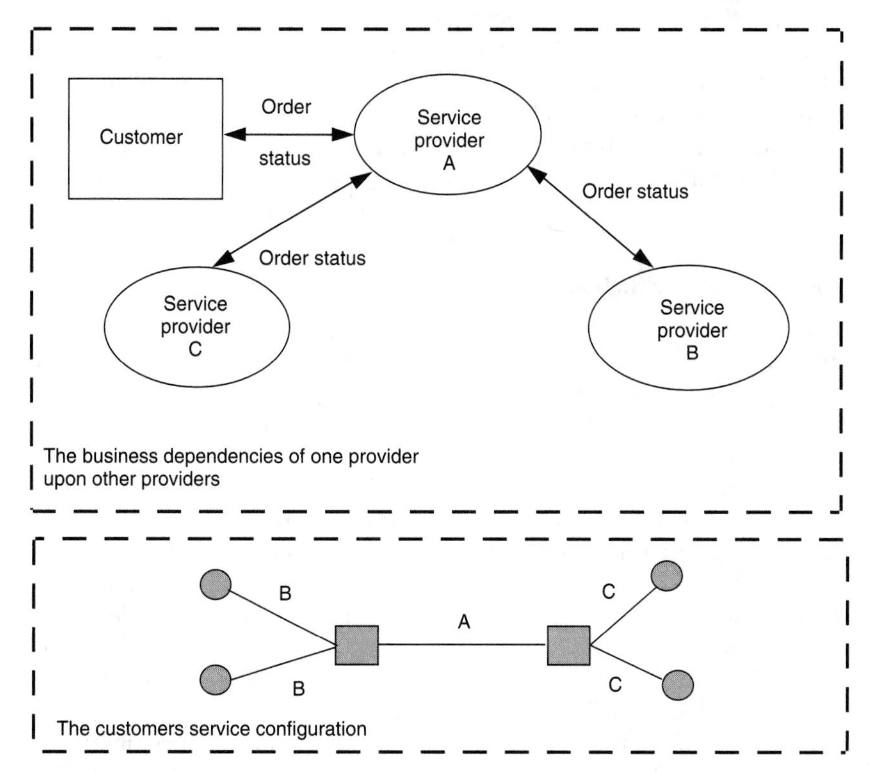

Figure 5.4 Service provider interdependencies can inhibit service guarantees.

Risk management

Any subcontractual relationship involves risk. The first provider in the chain, having made a commitment to deliver a service to its customer by a certain date, is dependent on its subcontracted providers to deliver on time as well. By integrating order-tracking processes and systems via an appropriate "boundary" interface, service provider A can track the progress of service providers B and C in completing their work, as shown in Figure 5.4. Without such process flow-through, service provider A will spend considerable time, money, and concern trying to reach people down the line to determine how the project is faring. With automated process flow-through, enabled by industry agreements, the risk of dependency on others in the service chain is reduced.

Speed of response

The second factor is the speed with which information is needed to respond to customer inquiries. To continue with the example shown in Figure 5.4, imagine that a problem occurs with the service. The customer contacts its service provider (A), which might discover that the problem really lies with the portion of service provided by service provider C. Initiating a trouble report or directing the initial report to service provider C must be done quickly, and action must be taken quickly by C in order for A to fulfill its service restoration goals with the customer. Relying on manual methods for exchanging trouble reports or tracking their resolution may be too slow to meet the rising expectations of customers in competitive markets.

Transaction cost reduction

The first two reasons for automated process flow-through are easy to understand—from the viewpoint of service provider A. But what's in it for providers B and C, aside from any interest they may have in taking care of provider A, which is paying them for their services and which may be a very large customer? The final reason for automating a provider-to-provider interface is important to both sides of the exchange. It has to do with operational costs. When the volumes of exchanges (orders, trouble reports, billing records, performance reports, etc.) exceed a certain threshold, manual information exchange has to give way to automated interfaces. Although the current U.S. model, in which every interexchange order spins off at least one and often two local orders, may be the extreme case, the increased alliance activity connected with building a global presence is starting to drive volumes up in other geographical regions as well.

In alliances, the partners may well be "friends" rather than competitors. Nevertheless, the trading relationship is usually still customer-to-supplier for any given transaction. Often the relationships are more complex, involving several parties that can be "customers" one minute and "suppliers" the next. Tracking of orders, billing records, and problems in this complex, dynamic environment makes service management automation and process flow-through an essential rather than optional approach.

Service Provider to Networking Technology Process Flow-Through

Some value-added service providers do not operate their own networks, so the links to the network management and element management layers of the TMN model are truly "external." Other more traditional service providers continue to operate their own networks, and it may be difficult for them to think of the interface to switches, network databases, or signaling systems as "external" since they probably own and maintain the networking equipment that makes up their networks. However, almost without exception, now that AT&T has split off its equipment business, service providers are reliant on third-party suppliers for their equipment. The provider's ability to configure the equipment, monitor its performance, recognize and respond to alarms, and collect usage data relies on the management capabilities of the equipment.

Integration of the information and control capabilities of element management systems with service and network management systems is critically dependent on appropriate, well-defined interfaces. As shown in Figure 5.5, the business objectives that are driving the automation of process flow-through across this link include faster service creation, easier (more standard) manipulation of the devices, and timely access to data that can be easily understood and processed by the service management systems.

The reliance of the service management layer on network and element management capabilities is significant. If a provider is to fully automate key service management processes, a link to these underlying systems must be established.

Faster service creation

The move toward advanced intelligent networking (AIN) is proof of the service providers' desire to gain more flexibility (by orders of magnitude) in the area of service creation. Although current networking technology is far more software-driven than the switches of old, even a minor change to a switch can involve many months of design and negotiation and a considerable waiting period to completion. The trend is toward moving as much intelligence as possible out of the switches and into distributed systems that can be modified more quickly and independently. The notion of specifying service "building blocks" implies a common set of functions or capabilities that can be used in different combinations to produce multiple service offerings.

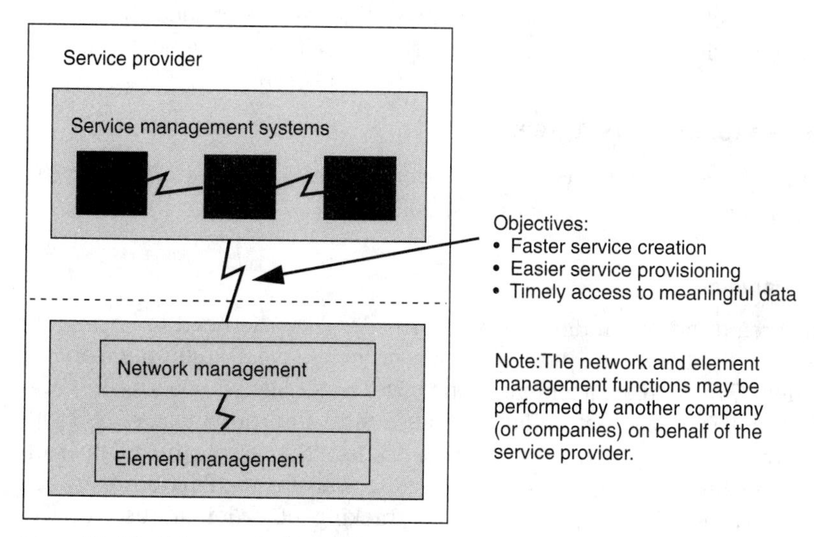

Figure 5.5 The link to networking technology.

Easier manipulation of supplier equipment

In addition to the one-time equipment modifications associated with service creation, day-to-day provisioning of service requires the ability of the service provider to configure (or reconfigure) the supplier's equipment. When there are multiple suppliers, each with its own way of receiving and interpreting commands via its own element management system, the job of linking service configuration and network configuration is made much more difficult. As service providers develop new services or improve existing ones, they need to reach agreement across suppliers to ease this task. Referring again to the example of adding features to a customer's switched service, if the same set of commands can be used (regardless of switch vendor) to add custom calling features to a customer's line record, that saves translation time, training time, and errors in the service provider's provisioning process.

Timely access to useful data

Making the link between data needed by technicians to maintain a device and data needed to manage a service is crucial and also very difficult to achieve. Until very recently, equipment suppliers determined what data to make available based only on the immediate user's (the technician's) needs to maintain the equipment. Now, service management requirements are pointing out the need for information that relates to the use of the equipment in a service or the impact of a failure on the service user. For example, a provider of virtual private services that does not own its own networks cited the desire to somehow "tag" physical facilities with the identification of customers to whom they are assigned, so that failures at the facility level could be quickly related to the customer's service level agreement (SLA).

Priorities for Industry Agreement

Judging by current industry actions, several key areas of interest to service providers are a first priority for process linkage.

Order handling

In the area of order handling, service providers see the need to be able to automatically track the status of an order between multiple service providers. The actual placement of the order is not likely to be a system-to-system interface, at least initially, because most interprovider orders tend to be complex, and, with the exception of the U.S. market, the volumes of order exchange are not sufficiently large to warrant tackling order transmission. However, by automating the tracking of order status, service providers believe they will considerably improve their visibility of service completion when another provider is involved and thus do a better job of managing any roadblocks to improve overall service delivery to customers.

Some service providers plan to extend any provider-to-provider automation agreements to their major customers, giving them a firsthand view of the status of pending orders and therefore increasing the customers' confidence in the service provider's ability to meet promised due dates.

One area where order placement is being automated is in the U.S. market, specifically to support Primary Interexchange Carrier (PIC) notification. This type of order is quite simple, involving very high volumes of transactions as customers switch from one primary long distance carrier to another, and local exchange companies are asked to carry out the changes. This work is being carried out within the Exchange Carrier Implementation Committee (ECIC) of the Alliance for Telecommunications Industry Solutions (ATIS), a nonprofit organization serving the North American telecom service provider industry.

Problem handling

Based on work done in 1991 by the Network Management Forum (NMF) and Committee T1 of ATIS in North America, service providers have already begun to implement agreements for exchanging trouble reports. Detailed agreements, worked out within ECIC, cover both the transmission of the trouble report and the exchange of tracking information and have been applied to the access circuits that make up most of the services ordered from local exchange carriers by interexchange carriers. Unfortunately, the implementation agreements—while based on a common set of underlying standards and specifications—are still bilateral, perhaps due to the fact that so much of the information required to populate a trouble report involves data held in a variety of legacy systems.

Going forward, service providers wish to extend the value of the work already done in this area. Their first priority is to reach global agreements (not on a bilateral basis) dealing with the exchange of trouble tracking information for specific service types. Leased circuits and toll-free switched services may be the first targets of their work.

Service providers also see value in building automated links with their customers' help desk systems, so that customers can access relevant status information through their own systems. This work has already begun and is discussed later in the book.

Performance reporting

Customers are starting to enter into service level agreements (SLAs) with their providers, and this competitive trend has opened many new problems for service providers and their customers.

Here are just a few examples:

1. Customers (particularly large and important ones) may negotiate individually tailored SLAs with their providers. This is because they, in turn, have SLAs with their own customers (the end users), and there is a desire to make these SLAs line up. While this is certainly understandable, it is extremely difficult for service providers to keep track of the many SLA variants that might be developed or to respond effectively to problems when the terms of a service might vary widely from one customer to the next.
2. On the other hand, customers who are not large enough to demand tailored SLAs are subjected to multiple SLAs from multiple service providers and other suppliers. They see some measures (such as availability) that may sound the same but represent very different measures. They see multiple terms for the same thing. With no standard terms or definitions, they are left having to make the translation (usually crudely) in order to produce some sort of measure of overall performance.
3. Service providers are apt to create SLAs around information that exists, whether or not it is meaningful to the customer. For years, providers have been reporting "errored seconds" to customers who probably just want to know whether their application was or was not affected. Ken Clamp, network services manager of National Westminster Bank's information technology unit, when describing his company's expectations of its service providers, said simply, "I need to feel that the provider takes my business seriously—that they recognize which of my services are most critical to me, that they nearly always deliver services in a way that is transparent to my end users, and that if a problem occurs, they demonstrate that they have taken all possible steps to keep my applications up and running. I don't need a lot of detail—just sensible business indications that the provider is on my side."
4. Finally, constructing an SLA is no good if it can't be tracked, and that is where equipment suppliers come into the picture. As service providers move to provide more meaningful measures of service performance to their customers, equipment suppliers may well be asked to provide network and element data that doesn't currently exist. If they receive multiple, conflicting requests from their customers (the service providers), they are likely to continue along their independent courses.

There are many benefits of agreements in the area of performance reporting. From the customer's perspective, the establishment of common terms and definitions for key performance measures would make it possible to construct an end-to-end performance view for their end users, incorporating not only services from the communications providers but the performance of any internally managed networks or devices.

From the service provider's perspective, having a common set of performance terms that could be used to construct service level agreements and could also be implemented at the network or element level would make it possible to link network performance to service level agreements automatically. Administration of SLAs would be simplified, and responsiveness to customers in the event of a failure would improve.

Billing

Service providers have been exchanging settlement payments for decades based on ITU-agreed formulas that divide revenues between originating and terminating locations. But the needs in this area are moving on, and service providers see areas where industry agreement and process automation could be extremely beneficial.

As customers of other providers, service providers want to see bills in common formats using common terms, just as large enterprise customers want the same thing from their multiple providers. This points to the need for a common usage record from equipment suppliers—an item that has been talked about for as many years as we can both remember.

In addition, some providers are outsourcing the expensive billing process, while others have made a business of providing billing services. Even some financial institutions are eyeing the bill preparation and collections business on behalf of service providers. When third-party billing agreements are struck, it is likely that those performing the billing function will mandate that certain formats be used in providing rated call data and account record information. And it will be to the service providers' advantage to make certain such agreements are accepted across the industry so that they have the latitude to change from one billing service company to another if circumstances warrant.

Service configuration

Common configuration agreements are needed that permit service providers to establish customer services or make changes to their configuration automatically. Say that a customer has a variable bandwidth service. For a fixed price equating to 15 data channels, the customer can expand its usage to a maximum of 30 channels, paying only for the increased usage when invoked. Making changes to this service requires that the underlying network elements (multiplexers or switches) be informed that the channels are to be made available and that charging is to commence. If this customer (or the operations support center) were required, each time a change was requested, to know which elements needed to be reconfigured and how to issue the appropriate commands to do so, the service wouldn't be very useful. Instead, automatic reconfiguration means that the customer can reconfigure the service in terms he or she understands, and the underlying network and element management systems will make the connection to the right elements and invoke the change. In a multivendor environment, automating this function can result in considerable cost savings and speed of enactment.

A simpler example of configuration is the application of calling features to a customer's residential or business voice services, as mentioned earlier. Adding three-way calling, for example, requires that a command be sent to the switch to activate the change. By establishing a common way to reconfigure the switch for this purpose, service providers will be able to manage service churn more quickly and with less manual intervention.

These configuration examples point out the need to make agreements as universal as possible across networking technologies. As shown in Figure 5.6, service providers would like a large percentage of management information to be common across all technologies. They acknowledge that some differences are to be expected between major categories of technology but are willing to accept only a small amount of "technology-unique" information, and only if there is a justifiable need.

Referring to the variable bandwidth example, it shouldn't matter whether the underlying networking technology consists of time division multiplexing (TDM), asynchronous transfer mode (ATM) switches, or synchronous digital hierarchy (SDH) technology. The commands transmitted from the service level to the network (and element) management levels need to be the same. If they are not the same, considerable automation benefits will be lost to the service provider.

Service monitoring and control

Similar to the configuration work, service providers want common ways to monitor alarms at the service level using information provided by the network and element levels of management. Again, these agreements are intended to transcend underlying technology type, giving the service provider

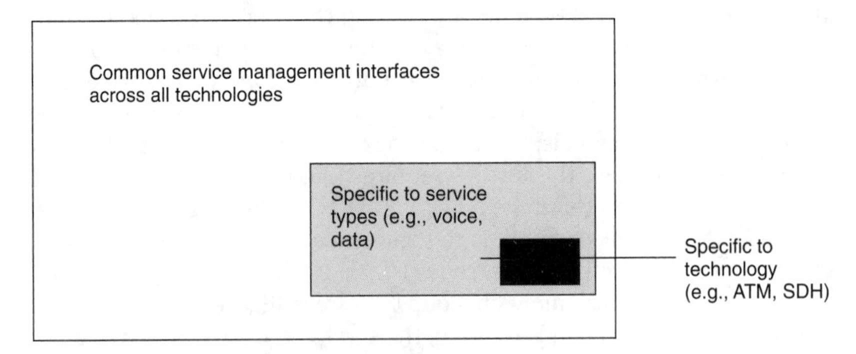

Figure 5.6 Common interfaces across technologies is a goal of service providers.

the ability to implement and monitor a service using the same service management systems regardless of what technology is used to create the end-to-end connection.

Interprocess links

Although service providers might wish to focus on improvements in one process or another, they should not forget that their many processes are linked, and they will need to be careful to maintain consistency between sets of agreements. For example, any agreements regarding performance reporting will need to be linked with agreements made in problem handling and service monitoring. To have one group of agreements reached on a set of SLA terms and a different set of agreements governing what information needs to be provided from the network monitoring to the service monitoring process will mean that end-to-end process flow-through can never be achieved.

This is a considerable challenge to the industry, particularly at a time when the sense of urgency is pushing providers to solve specific problems quickly. Service providers and their suppliers will do well to keep an eye on this potential problem as the various industry groups set about meeting the needs that have just been articulated.

Why Not Just Create Bilateral Agreements?

In all three types of interface links (with customers, other providers, and suppliers' network and element management systems) the need for automation has been stressed. But automation can occur any time two parties reach a mutual agreement. Why is it necessary to implement agreements in common across the industry?

Service providers give several reasons why they prefer to implement industry agreements over proprietary ones at the crucial "boundary points."

1. Without a common way of "viewing" all of the elements in an end-to-end process, integration is virtually impossible. A series of different proprietary interfaces would create great barriers to fully automated process flow-through.
2. Considerable time is required to negotiate one-on-one agreements, and some compromise is generally involved. Reaching and then maintaining multiple one-on-one agreements is costly and inefficient, and in the case of agreements with customers, simply impossible because of the number of customers involved.
3. Providers may have multiple equipment suppliers as well as multiple sources of subcontracted services. If a few crucial interfaces can be

made common across all suppliers, the provider has a chance to deliver end-to-end services faster and more efficiently.

4. Customers want a common approach from their suppliers that lines up with their customer premises systems. They do not want the expense and inconvenience of multiple proprietary interfaces from their service providers.
5. The convergence of the media, cable, computing, and telecommunications industries to provide value-added services suggests that the ability of a provider to connect with others will be key to its future success. Being able to move swiftly, to say "yes, we support a standard interface" is becoming an increasingly important commercial issue.

Chapter

6

Service Management Perspectives of Private Enterprise Network Operators

While we have aimed this book at companies that provide communications services as a business, the need to achieve service management excellence applies equally to internal managers of private enterprise networks. Nevertheless, the detailed challenges faced by each are different, and this section is an overview of some of the more specialized service management requirements of private enterprise network operators.

Service managers in major corporations are generally part of the information systems (IS) or technology (IT) department. The IT service manager of a global hotel chain, for example, is responsible to see that the reservations center can gain access to room availability records, that accounting data is not lost between an individual hotel and the corporate accounting system, and that employees in the marketing and sales division can access electronic mail without difficulty.

The term *service management*, as used by these managers, means meeting the requirements of their internal end users. To fulfill that responsibility, the corporation's service manager relies on many different types of suppliers, including:

- Suppliers of services (communications and information service providers).
- Suppliers of equipment (computers, premises-based switches, modems, hubs, routers, printers, multiplexers, etc.).

- Suppliers of applications such as electronic mail, office systems, and database systems.

Just as the communications service providers are now entering into service level agreements with their customers, enterprise service managers are bound to meet SLAs with their various internal clients. They must demonstrate continuously that the service they provide is at least as good as what could be achieved through outsourcing and that the cost they represent to the corporation is competitive with what a systems integrator or other management company might charge.

A Process Model for Service Management in an Enterprise Environment

Some time ago, the members of the User Advisory Council of the Network Management Forum outlined, for the first time, the sets of processes used to meet their clients' service level expectations. Figure 6.1 summarizes these processes.

Although different companies may describe these processes using different terms, the functions outlined below represent the consensus view of the service managers of 20 major enterprises representing the financial, airline, and parcel-delivery service industries, as well as manufacturing companies. These processes are:

- Client contact (which includes order handling, problem handling, and inquiry handling)
- Fault tracking (which includes tracking, escalation, and progression of problem resolution against service-level agreements)
- Change control (which tracks orders, routes orders, and provides overall project management of major changes)
- Performance monitoring (which manages usage statistics, analyzes faults, monitors SLAs, and tunes performance of the networked information system)
- Planning and design (which includes capacity planning as well as the design, planning, and tracking of major changes)
- Implementation and maintenance (including configuration management, implementation, and provisioning management)
- System administration (including the management of software applications, computer systems hardware and software, and end-user configuration)
- Fault monitoring (which monitors the networked information system for potential service-affecting problems)

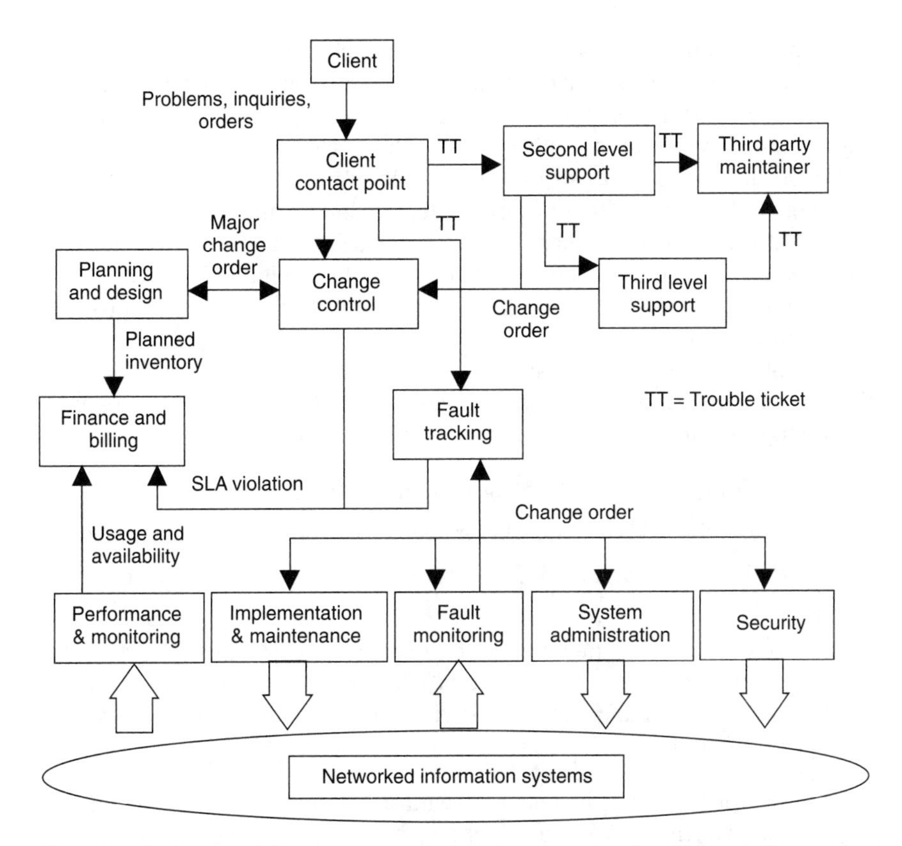

Figure 6.1 Business model for managing private enterprise networks. *(Network Management Forum User Advisory Council. 1992. Statement of User Requirements for Managing Networked Information Systems.)*

- Finance and billing (including calculating and apportioning costs, asset management, billing, and bill reconciliation)
- Security (including administration, detection, and recovery)

As in the case of the communications service provider, continually "raising the game" in service management by automating the end-to-end flow of these processes is of crucial importance to enterprise managers, who face similar pressures to reduce costs and improve customer service quality. But the needs of enterprise managers are somewhat different from those of service providers when it comes to automated flow-through priorities.

Enterprise service managers would like to purchase applications to help them manage their networked information systems. They are looking for third-party applications that will help them manage individual processes such as change control or trouble reporting and tracking in such a way that the individual applications interoperate to support end-to-end process automation. They are less concerned about "external" interfaces, perhaps because their "customers" are individual end users (not organizations with management systems), and interfaces with peer companies (e.g., bank to bank) regarding the status of either company's communications network are likely to be very limited. For example, two banks may exchange a great deal of information about financial transactions (for which banking industry agreements have no doubt been reached), but they might only exchange information about their networks in the event that the communications link between the banks is faulty.

Most of the concerns of a service manager of a global private network, then, are focused inward. He or she might well ask the following questions pertaining to managing the delivery of services to end users:

- Am I meeting my service level agreements with my clients?
- Can I respond quickly to requests for new service or for additional capacity?
- Do I deliver good value to the corporation, as measured against outsourcing benchmarks?
- Am I viewed as a key member of the executive team in finding creative, cost-effective ways to use technology in order to deliver new services (e.g., banking or transportation services) or new products (e.g., cars or semiconductors)?

The enterprise service manager's organization might consider the following questions at the network and systems management level:

- Can we use public facilities to handle accounting's nightly bursts and payroll's monthly surges?
- Are our servers and applications properly configured for maximum business effectiveness?
- Can we reassign users to reduce printer congestion?
- Are we able to manage the upgrade to new software without disrupting service?

Integrating Multiple Computing Domains

If the service management levels of an enterprise depend on the ability to seamlessly move information between a variety of systems, it is important that the enterprise manager be able to integrate several computing domains. Figure 6.2 is one enterprise manager's depiction of the enterprise management environment.

When trying to meet service level agreements with end users, the main challenge to the enterprise manager is coordinating among the desktop,

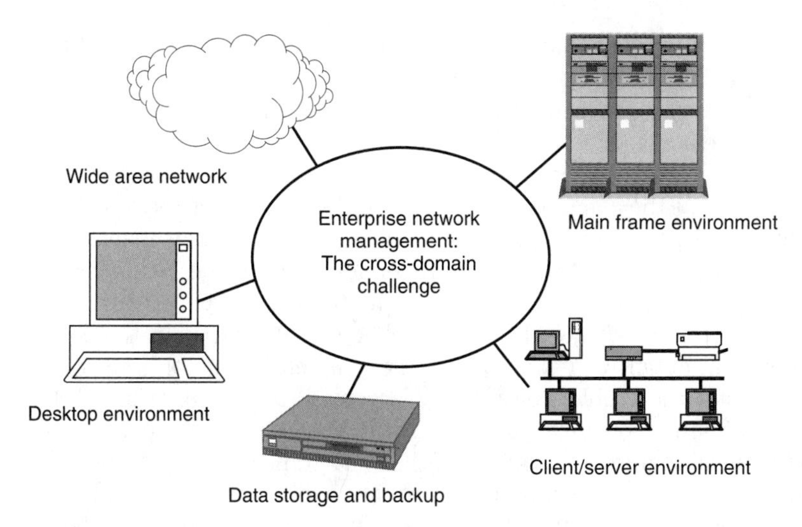

Figure 6.2 The management challenge for enterprise network managers.

client-server, and mainframe environments, and dealing with the storage, access, and backup of critical data. As individual departments make more and more purchasing decisions at the desktop and client/server level, tying these domains together has become increasingly difficult.

It is no wonder, then, that enterprise managers are anxious to be able to integrate the management of networks and services across these different domains. Consider these two broad categories of service management applications needed by the enterprise manager:

1. User administration applications.
   ~Password administration across multiple business applications.
   ~Add/delete/change users by entering data once to update multiple systems.
   ~Establish and track SLAs and report performance to SLAs.
   ~Client contact "help desk" to handle problems, take orders, answer inquiries.

2. Systems management applications.
   ~Backup and recovery across multiple domains.
   ~Common data repositories to be accessed by any of the computing domains.
   ~Capacity planning and allocation across domains.
   ~Automatic fault notification (using common terms across domains).
   ~Performance reporting (using common terms across domains).

Traditionally, the computing industry has approached management differently from the communications providers. Whereas the telecom service providers have viewed management from a total business perspective (through the TMN model), systems management solutions have evolved in an ad hoc manner for individual domains. Cross-domain integration is left to the enterprise manager.

Over the years, service managers of private enterprise networks have tried to influence the computing suppliers to address their cross-domain integration needs, with little positive effect. Forcing product changes requires a concerted, focused, long-term commitment on the part of users, which most enterprise service managers simply can't make. It is very difficult for enterprise managers from different types of companies to discuss their needs in detail with each other (the competitive factor comes into play, as well as the problem of industry jargon). It is also very difficult for those with line responsibility to devote sufficient time to developing detailed technical requirements—even if they had sufficient expertise.

Two industry sectors are attempting to make a difference by pooling their collective buying power. One is the petrochemical industry, and the other is the telecom service providers—this time in their role as very large enterprise users.

In fact, the telecom operators represent the largest users of computing equipment in the world. People don't often think about the IT side of a service provider, so they overlook the potential this group brings to helping sort the problems of multiple-domain IT management. Information-systems people within service provider organizations share the same problems as the enterprise service manager described in this chapter, except that the scale of problems is generally larger than most other enterprises.

This industry-focused group of enterprise managers (the IT managers of the service provider companies) has both the expertise and the motivation to cause crucial changes to be made. Later in this book, we highlight their efforts as we review the NMF's SPIRIT program. We encourage private enterprise service managers to take note of their work, and, even more important, to take advantage of their results.

Part

2

It's All about Integration

So far, we have focused on the business motivations driving service management excellence, with views on where process linkage is most important. In this part of the book, we discuss what is involved in trying to integrate and automate management process flows.

Reengineering a company's service management capabilities is among the greatest systems integration challenges in the industry today. Perhaps this is because management, by its very name, implies the ability to correctly read a situation and to take the right actions—quite a different undertaking than simply filling out a form or computing a bill. The amount of integration required to achieve end-to-end process automation—between management levels and across the service delivery chain—presents a challenge that few systems designers have ever faced.

Service providers that emerge from the inevitable shakeout of the industry over the next decade will be those that realize that end-to-end process automation is a basic survival issue, not an option. Yet that task must be undertaken with the greatest of care, because the implications of failure are enormous. One difficulty with ever-advancing communications technology is that a single error has greater and greater potential to affect multiple customers. Back in the days of the Strowger switch (you might know it as step-by-step), a single failure meant a single customer's service was affected. With the advent of common-control electronic switches, a single failure could mean an entire switch (probably 30,000 customers on average) could be affected. As switches have been integrated into hierarchical networks, a single failure can now affect hundreds of thousands of customers at one time. A similar set of risks occurs when management processes move from being based around people, to stand-alone management systems, to a fully automated and integrated environment.

The management of such a complex, interconnected networking environment is nontrivial. And as service providers move to integrate across the element, network, and service management levels, as well as across the service delivery chain, complexities grow. In this part, we examine the need for integration at both the process and systems levels and the barriers that make that difficult today.

Chapter

7

What Do We Mean by Integration?

We've already introduced terms like automation, integration, and process flow-through. But what do we actually mean?

Our objective is a highly efficient service provider infrastructure in which the business processes are able to be conducted with a minimum of human intervention. In essence, we are advocating a revolution in the way communications companies are run that is as fundamental as moving from yesterday's networking world of switchboards and operators to today's fully automated digital highways.

Process flow-through—the seamless flow of information about orders, problems, changes, accounting, and so on—needs to occur both within a provider and across the entire service delivery chain as necessary. To do this as effectively and efficiently as possible, several things must happen.

First, the process steps must be automated so that those tasks within the process that are done by people can be done by a computer system. Second, the systems that support the various process steps must be integrated together in some way that doesn't require people to be the "glue" that holds the process together.

But just automating and integrating existing processes isn't going to get the organization to lean provider status. Instead of simply mechanizing processes that were developed for a manual world, the processes themselves must be rebuilt (reengineered) to get the full benefit of automation and integration. That doesn't mean that all of the legacy systems within a service provider are instantly out of date. It is possible to progressively move to the lean state by defining and implementing process flow-through interfaces at key points between systems that make up the overall process. These systems may be quite dissimilar, especially if the interface is between two different organizations. But flow-through can be achieved, provided that the information that passes between the systems is rich enough to obviate any human intervention.

Achieving automated process flow-through means that integration must have occurred at two levels—integration of the business and management applications that make up the process from a "work" perspective (order handling, billing, etc.) and integration of the underlying systems technology so that data can pass seamlessly across multiple infrastructures.

An integration example

Perhaps an example of information flows in a human context will help make the point. Two communities develop independently for many years, one a farming community and one a fishing community. Each creates its own customs and its own rules of acceptable behavior. At some point, they decide that both could benefit from doing business together—exchanging fish for vegetables so that everyone can eat a balanced meal. But although they choose to agree, achieving the goal is quite complex. For starters, they don't speak a common language, so a few basic terms need to be established and understood—terms like "how much?" They don't have a common currency of exchange. They don't conduct transactions in the same way. If they are going to do business together, they need to figure out how to place a value on each other's goods, whether they will transport goods to each other or make use of a central exchange point, and so on. If they accomplish their objective, they will have integrated their communities for the purpose of exchange, even though both communities remain unique in their language and customs.

Processes and systems are similar to humans in that they are all different for a reason. Bringing them together for a specific business purpose requires that in that one area they agree to do things the same way. Initially these areas of agreement will tend to be fairly limited in number, but they can grow. The farmers and fishermen might decide after a year or two of successful cross-marketing that they would now like to join forces to defend themselves against the mountain village people, requiring more advanced agreements and a much closer cultural tie between the two communities. Similarly, agreements across processes and systems are likely to start small and then expand as the benefits of integration are realized.

Objectives of integration

There are many business advantages of integrating processes and systems, and we have touched on some of these already. A simple agreement on the use of common terms (integration of information) can result in improvements in accuracy as information is passed from one step in the process to the next, even if the exchange is performed via paper handoff. An example is reaching agreement on performance measures that are meaningful to customers.

Going further, agreeing that common information will be exchanged automatically, from one system to the next, brings reductions in manual data entry. This means further improvements in accuracy as well as increased speed, even if the systems used at each process step continue to operate as largely independent islands connected by a thin bridge. Expanding that thin bridge into a superhighway, perhaps by establishing a common database accessed by both systems, starts to take costs out of the systems themselves and makes it easier to extract information that can be used by new applications to enhance service capabilities offered to the customer.

Going still further, if both systems were built on platforms that had similar characteristics—a common user interface, for example—it becomes possible for people accustomed to operating one system to learn the next more quickly, and individuals who might be required to use both systems as part of their jobs will make fewer mistakes when the command structures are similar between the two applications. Managing the development of systems becomes easier and less costly when integration extends to things such as programming languages or application programming interfaces (APIs). Managing the operation of the systems becomes easier when both provide similar monitoring information, using the same terms.

Automated Process Flow-Through

We don't think for a minute that it is possible to automate every part of every service management process or that such a goal is desirable. Human beings are much better at interacting with other humans than computers (at least with today's technology). Thus it is important to retain people at the customer interface, and we would caution against being too creative here. But we have learned the hard way that it is not a good practice to rely on people to make crucial, time-sensitive, or interlinked processes work. Rely on your people to pick up the pieces when the systems inevitably fail, but don't rely on them as the means of achieving integration if your organization wants to be the industry benchmark for service management excellence.

Why manual improvements are not enough

While it is true that there is more to managing service than getting systems to talk with each other, the environment most service providers face today calls for automating process linkages in order to reduce costs and errors, give better service, and keep up with the pace of change. The challenges of automated process integration vary, as do the consequences of failure, as the following stories illustrate.

Illustration 1: The telco shuffle. We have had some good chuckles when comparing different service providers' efforts to deal with internal shuffling. One of us remembers a concerted, company-wide campaign that promised we would be "easy to do business with." The goal of the campaign was to stop sending a customer from one department to the next, and the next. Every employee was given a directory of internal telephone numbers in order to improve the chances that the customer might be cared for after only one (not several) call transfers. In essence, this was an attempt to improve service by modifying manual processes. It cost a lot of money to compile and print the directories, to train employees, and to promote the program internally. But instead of improving service to the customer, it had the effect of increasing the number of call transfers. Relying on people to make the right connection between a customer request and a departmental function spelled disaster as each person made different choices with each new customer call. The result for customers? Inconsistency, delays, and confusion. The program was short-lived for good reason, but it illustrates one of the primary reasons for automating and integrating internal processes: to make it seem like the left hand knows what the right hand is doing—consistently!

Illustration 2: Betting that you can deliver. Next in line in the annoyance category after the telco shuffle is not being able to ascertain when a requested change or new order might be completed. In the United States, for example, try ordering an ISDN line from the local exchange carrier. (It doesn't matter which one, at least based on our recent experiences.) First, it takes a long time to find someone able to accept such an order, and the customer must provide the technical detail needed to configure the circuit. Getting an installation date is not possible—it happens when (and if) it happens, and there are no guarantees.

Presumably this is because the many different groups involved in provisioning an ISDN service are not able to exchange information. In other words, the provisioning process for this service is entirely manual, and there is no way to guarantee a date at which an order put in at one end will result in service at the other end. Sadly, many telecom service providers operate on this same principle even with simple services. When processes are not automated—when people must make the linkages through the process flow—it is not possible to quote a firm due date. Where dates are quoted based on historical performance (or a service objective) without the systems to back them up, actual performance will vary widely. And telling a customer that service will be in "no later than" a certain date is no real help, either. Customers are put in the position to wonder whether the service is in or to call and find out (and go through the shuffle again).

Automation is crucial to achieving service-level consistency, and it is important in every phase of service management. Service consistency means the ability to confidently quote due dates, to state that a change to the billing record has been made while speaking with the customer, to apply an immediate rebate, or to provide accurate and up-to-the-minute status information regarding a reported problem. Service consistency requires automation (and integration) of management processes.

Illustration 3: The road to service outage is paved with good intentions. Illustrations 1 and 2 are fairly easy to fix compared with this one, which begins to involve real-time service operation. This actual experience mirrors many we have had over the years. It shows how, on the one hand, sophisticated management systems integration is critically important, but, on the other hand, just how hard it can be to do right.

Within a service provider's operation, two management systems—a network management system and a service management system—are tied together at a primitive level (exchanging status information) but mainly rely on humans to make the link between them. Both are distributed, as are the databases, and one is physically located in an unmanned site. Maintenance of the hardware and software below the management application is under contract to a third-party computer company, which needs to complete a software upgrade to the operating system. All goes well with the upgrade and the system is restored, except that there is now a mismatch between what the network management system is reporting and what exists in the customer's network.

Because this is a multidomain network, there is not a single network view even at the element level, and the different domains have to be managed separately and manually. The service manager starts to see things go wrong in the customer's network and determines that it is the element management system misreporting. He tries to reset the software and is permitted to do things that shouldn't have been allowed (because the software isn't fail-safe). Had the software been left alone to recover, it would have sorted out the problem, but instead the technician decides the management system has failed and reboots it right in the middle of its attempt to reconcile databases. Now, nothing is right any more, and when the network actually does fail, the technician can't see the failure, nor can he get it back up and running once the problem is reported.

Several integration problems are illustrated by this horror story. First, the level of integration between the two systems was not sufficiently sophisticated to allow either system to make sense of what was happening. Second, the software wasn't human-proof (but then it rarely is). Third, the technician was not adequately trained. Two different departments were involved in integrating the systems. One was given instructions on what to do (or not to do) in the event of such an occurrence, while the other was not.

Two significant points stand out from this chain of events. First, it shows the immaturity of the supply chain linkage—the inability to properly relate changes in an underlying computing system to the networking equipment and management applications that make use of its capabilities. Second, it shows just how sophisticated a really good service management capability needs to be in order to react and respond correctly to problems that will invariably happen.

The pain and pleasure of pursuing automated process flow-through

The methods used and the rate at which different companies move to automate and integrate their processes and systems will vary, but two things seem to be common: First, it is very hard to achieve the first success, and second, once the trend is begun, it usually continues.

As Wendy Shorrock, an executive at BT responsible for network management collaboration, told us, "The first steps to make things better often complicate rather than simplify processes and can lengthen rather than shorten a process." As one who has experienced the pains as well as the benefits of making significant changes, Shorrock advises service providers to be prepared to invest a lot and have patience to see the benefits of integrating their management processes and automating process flow-through, because it's not as easy as it looks.

But the benefits are real once achieved, and companies that can measure the value of integration in terms of process flow-through improvements or the cost and time to develop new applications generally continue along that path until they have fully exploited its value.

Process-Level and Systems-Level Integration

The key to achieving service management excellence is automated process flow-through, and the key to achieving that is process-level and systems-level integration.

Process-level integration has to do with reaching agreements on the information that will pass from one function to another and the expected behaviors of each function. In a mechanized world, these functions are often performed by software applications that are in turn supported by a computing systems environment. For the process-level applications to be integrated, it is also necessary to achieve integration of the underlying systems. Systems-level integration has to do with making it possible for information to be exchanged among systems, even if one is a mainframe, one a UNIX server, and one a personal computer.

Process-level integration occurs within lean service provider companies both internally and externally. Internally, each provider is striving to improve process linkages to beat the competition, and this work is often highly proprietary. But to achieve true end-to-end process flow-through, most service providers must also examine external process linkages with the processes of other companies, including customers, service providers, and suppliers. In those areas where service providers see the need to achieve process linkages outside their own boundaries, they will seek agreement with other organizations. Otherwise, they keep their work close to the vest!

Systems-level integration is needed to support both internal and external process linkages. In an environment where customer data is spread across multiple mainframe systems, where network inventories are stored in large servers, and where operations center access to applications is via PCs, it is not enough just to reach agreements at the process level. It is simply not possible to achieve process linkage unless the underlying systems that contain the data can be linked (and managed) effectively.

In the next two chapters, we take a more detailed look at these two types of integration needs.

Chapter

8

Process-Level Integration

Integration of management processes needs to be viewed from two perspectives: horizontally and vertically (Figure 8.1). These two perspectives have some interesting ramifications when process-level integration is to be supported by systems applications.

Horizontal Integration

Horizontal integration enables interaction to take place between systems at the same management level. Within the service level, for example, integration allows a trouble management application to exchange customer data with a performance management application. It allows performance trends to be input to the planning-simulation tool. Many of the intercompany integration requirements identified by service providers involve this type of peer-to-peer interaction. Exchanging trouble report status or the progress of an order are examples of service-level horizontal integration across company lines. At the network and systems management levels, peer-to-peer interaction enables software distribution applications and license management applications to share the same user information, and network assignment systems can be connected with equipment inventory systems.

Partitioning of service management information

As in human interactions that are peer-to-peer, the information used by one process or system can usually be understood by another at the same level. It is likely that when doctors get together at their annual convention, they can

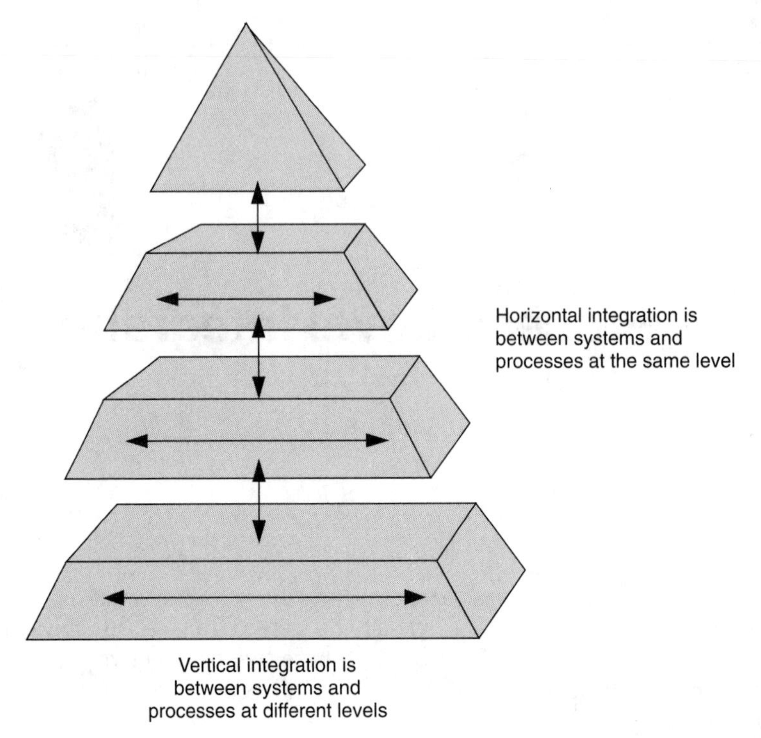

Figure 8.1 Process-level integration must be viewed both horizontally and vertically.

all relate to the same kind of information, and it's not necessary for one doctor to use layman's terms to explain a breakthrough to a colleague. What is necessary, however, is the ability to determine which information should and should not be exchanged, with whom. One doctor might choose to tell his colleague only enough about his work to get an answer to a troubling question but not enough to risk having the colleague steal his research idea. In the systems world, this is called partitioning, and it has interesting uses in service management.

Partitioning is an increasingly important concept as management personnel and systems are distributed across departments and around the world. Partitioning is really a way to view data access, and it is the principal business reason for requiring system security. Here are a few examples of how partitioning is used:

By time of day or day of week. A global service provider or consortium of companies might easily transfer coverage of its network from London to San Francisco to Tokyo as the work day progresses. A financial-services company might consolidate its service management functions to a single location during the weekend when demands on the network are fewer.

Depending on the day or time, access to service or network management information is permitted to some personnel and denied to others.

By location. The ability to allocate management responsibility across multiple work locations is a very effective way to manage productivity and to keep operational costs down. In a widely distributed network, it is possible to permit branch offices or work centers to access certain data or to perform certain functions while reserving other functions to be performed only by a central group. For example, a company might establish payment processing centers in each of the countries in which it operates so that customers do not have to pay international postage rates. Although customer payment records from all countries need to be visible to the order handling process, only the designated payment processing center is given the ability to update the information and only for the designated customers whose accounts fall within the processing center's jurisdiction.

By service or customer type. Among the things that differentiate service offerings is the level of operational support provided to customers. Very large customers or customers who make use of advanced, value-added services, for example, might be supported by dedicated work centers that are well-versed in the service offering or the customers' particular configurations. The ability to give a certain work center access to specific customer data gives the service provider added flexibility in differentiating its service offerings. This type of partitioning is also very effective when analyzing market sectors.

To support an alliance. As service providers enter into a variety of relationships with other providers, they will want to support different levels of data access. Close partners with whom the service provider has a financial incentive to serve as coprovider to an end customer might be given almost unlimited access to management data. Arm's-length relationships, on the other hand, will probably involve exchanging only minimal data in a tightly controlled environment to prevent competitive information from crossing the line. It's one thing to give a competitive service provider visibility of the status of trouble reports that directly pertain to its services but quite another to permit free access to the trouble administration system or the database of pending trouble reports.

Vertical Integration

The TMN model describes the different layers of management used by service providers in their business. Each of the four layers of management—business management, service management, network and systems management, and element management—is unique in its objectives but needs to be linked to achieve service management excellence. Linking of these layers requires vertical integration.

Vertical integration: An example

The service manager of a manufacturing company asks her network service provider to reconfigure her company's wide-area service to increase the bandwidth between London and New York to support a video conference. At the service level, the provider determines that the customer is authorized to make this change, and a command is initiated to the network level.

At the network level, a check is made that capacity is available, and a route is assigned. The network level then instructs the various element management systems to set up the physical link. When that is completed, each element-level manager confirms completion to the network level, where reserved assignments are taken out of available inventory.

The network level confirms completion to the service level, which alerts the billing process, notes the change in the customer administration and service-level monitoring systems, and finally notifies the customer.

What might have taken several weeks using manual processes can be accomplished in near-real time using advanced service management techniques. This is a real example of the lean service operator concept in practice. Advanced service management flow-through techniques enable new and improved customer capabilities, faster and cheaper than nonintegrated, manual methods.

The challenges of vertical integration stem from the fact that different management levels require different levels of detail to do their jobs. A key systems concept called data abstraction is central to successful vertical integration.

Data abstraction

The concept of abstraction has to do with varying the levels of detail of management information provided to each work process involved in service delivery. Said another way, it is a way to view the same data from different perspectives, depending on the functional needs of each user of the information.

For example, in an intelligent network, a service control point might malfunction in some way, and the element management system generates information that is sufficiently detailed to enable technicians to identify the problem and correct it. That same information, abstracted to a higher level, can be used to alert the network manager that the failure has caused a loss of connectivity on a certain route so that he or she is able to reroute traffic. At the service level, a further abstraction identifies that a specific customer or a set of users of a specific service might have experienced delays or interruptions (Figure 8.2).

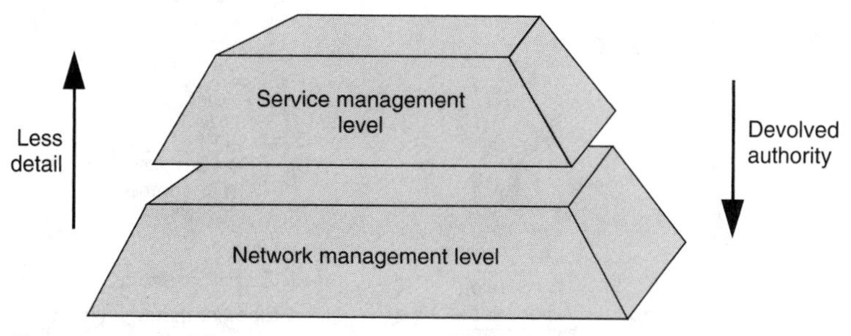

Figure 8.2 A crucial management function is to abstract data to meet multiple needs.

Similarly, when a customer orders a service, a great deal of information about the customer's requirements is needed at the service level. At the network level, it may only be necessary to know that additional capacity is required or that new facilities must be installed. At the element level, it is likely that customer-specific information may not ever be necessary, although the cumulative effect of multiple customer orders may result in the need to increase capacity.

This concept is very similar to human levels of management. Senior-level managers only get alerted if share price-affecting issues come up; otherwise they devolve detailed decision-making to subordinates.

Recursion: The Complexities of the Service Delivery Chain

The complexities of the service delivery chain have special implications for system design, because each point in the chain views its outputs as a service provided to a customer and its inputs as elements needed to create that service. To an enterprise customer, the ultimate service is what is delivered to internal end users, and communications capabilities ordered from service providers are but one element to be used in creating that service. A service provider that looks to other providers for local access connections, for example, views access service as but one element of the service it sells to customers. The provider of access service, however, considers its output to be a service and views the network services it receives from network operators to be elements, and so on, as shown in Figure 8.3.

If end-to-end service management flow-through is to be achieved, it is not enough for agreements to be reached exclusively at any one part of the service delivery chain. Only when information can be understood and manipulated as it moves from one link in the chain to the next will maximum value from automation be achieved.

Again, this concept has human management parallels. One person's boss is another person's subordinate. Each management hierarchy is recursive with the next. This is what value chains are all about. The CEO of a com-

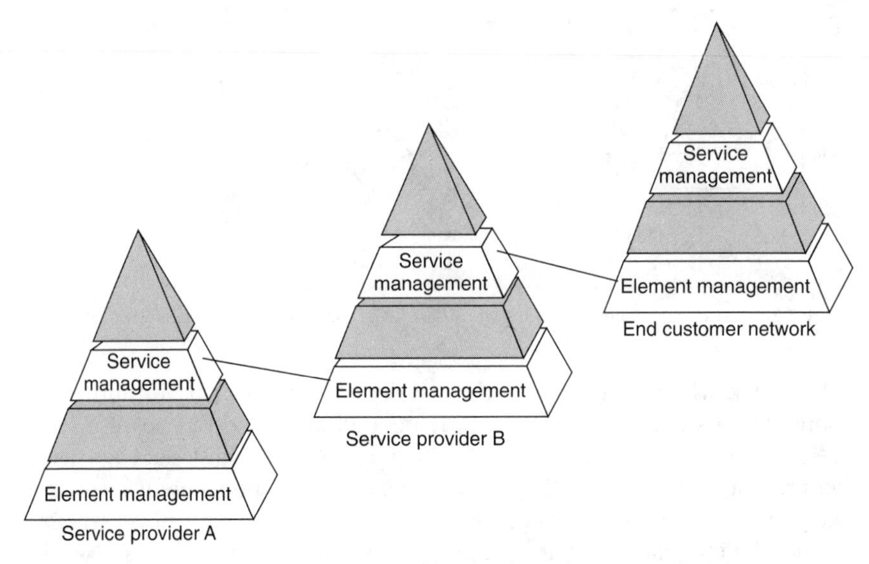

Figure 8.3 In the service chain, one company's service is another company's element.

pany can be in complete control of his or her own domain yet be at the bottom of the pecking order with his or her customer.

The Systems Are the Processes

The concepts just reviewed—vertical and horizontal integration, partitioning, abstraction, and recursion—add complexities that need to be considered when reengineering the business processes of service providers. First and foremost, they signal a need to understand the entire set of business processes that are affected when any one change might be made. Second, they present challenging requirements for managing the security of access to information. Finally, the information itself needs to be constructed with an eye to its possible uses by multiple processes and by multiple entities.

Let's be very clear about what we mean by end-to-end process automation. We don't mean perpetuating the building of systems whose applications merely automate today's human functions. That just leads to a faster version of an inefficient process. We mean that the integration of information passing between systems becomes the process, automated and integrated end-to-end. This is a profound change of viewpoint for many people used to designing a system specification that "fits in" with a people-oriented process. In the lean service operator model people exist at either end of a process (one touch!) but not in the middle of its execution. Process redesign can therefore be radical, significantly reducing the number of steps required. Rather than letting outmoded processes dictate systems that mimic human behavior, the systems become the processes.

This might or might not be good news, depending on the state of the legacy systems within a provider. Some companies might be able to move to the lean provider state mainly by integrating their existing systems. Others, and there are an increasing number who take this view, believe that simply integrating an inefficient or ineffective set of systems will not give the quantum leap forward in service management that they require. This group is looking to radically overhaul and rebuild large parts of their service management infrastructure and have begun to take significant reserves against earnings to provide the funding for this bold but risky approach.

These two approaches—evolve or rebuild—have pros and cons that each operator must weigh for itself. The jury is currently out on which approach offers the best return on investment. The rebuild camp does have some significant advantages. They truly can design systems that become a highly optimized process chain as opposed to integrating systems that themselves were usually produced to replicate less-efficient human processes.

Chapter

9

Systems-Level Integration

In process-level integration, the emphasis is on exchanging information in a common way. The emphasis of systems-level integration is on facilitating that exchange across computing environments that probably were not designed to work together. When each system used in a process has its own way of storing data, its own data formats, its own methods for permitting access, and so on, creating links can be extremely difficult.

Integration within and across Computing Domains

There are three main computing domains that must be considered in systems-level integration: the mainframe, the desktop, and the mid-tier domains. Each has developed independently of the other, and different computing equipment and software vendors have driven forward progress. The result is that there is little integration within each domain and even less between domains.

The glass house: Mainframe computers

The mainframe world is the most established and embedded of the three. It used to be that every computing application was developed on a mainframe computer because that was the only option. Now, the mainframe environment is selected less frequently, but it still makes sense for applications that have very large-scale processing requirements. As the most mature of the computing technologies, the mainframe environment offers good systems management tools and good security features, among other things. But the advantages of this environment have corresponding disadvantages, as well.

Because mainframes have large processing capacity, they are typically used for very large applications with millions of lines of code. Making changes to these massive programs is tedious, risky, and extremely time-consuming. And the richness of manageability that comes with proprietary mainframe systems brings the corresponding disadvantage of any nonstandard environment—an inability for mainframe-based systems to interoperate in an open way with systems that are built upon any other computing environment. Indeed, even within the mainframe domain, different vendors' computing equipment and operating software do not interoperate in an open way.

The desktop: More proprietary systems, more challenges to integration

Few companies could do business today without the personal computer, and the productivity gains it has meant for most office workers are significant. However, the effort needed to maintain and administer whole fleets of desktop devices and their related local area networks tends to cancel out any staff reductions that might otherwise have been realized.

The desktop environment, like the mainframe, has grown up as a proprietary domain. Over the years, fights about whether the PC or the Mac is better have grown stale (or irrelevant). Of more interest is that in the PC world, at least, manufacturing of the "boxes" has become a multivendor, diversified business, while development of the operating software, as well as an extensive array of applications, has been consolidating. In the space of a very few years, Microsoft has created a base of 80 million customers who may jump from IBM to Dell to Compaq to Gateway for their hardware but who tend to use Microsoft at least at the operating system level.

From an integration standpoint, this is good news and bad news. The good news is that with a single supplier of software at the desktop level, integration across the vendors' hardware within the domain is not an issue. The bad news is that a single, dominant software vendor might not always see business advantages in integration with other vendors—particularly those that supply the mid-tier and mainframe domains that still represent a huge legacy investment for most large companies, especially service providers.

Mid-tier systems: Filling the gap

Even before the desktop became the force it is today, a mid-tier computing domain began to take hold as "minis" and "maxis" gave developers the power they needed for most applications at much less cost and space than a mainframe. The operating system of these mid-range systems, typically UNIX, was initially developed not by a computing company but by AT&T Bell Labs. Perhaps that's why the mid-tier got an early foothold as the systems environment of choice for developing network peripheral applications, service management systems, and other new service management applications.

But the development of robust, manageable mid-tier systems has been dogged by a range of problems. First, as a backlash from too many years of dealing with the downside of a closed, proprietary mainframe environment, users of these mid-tier systems demanded openness across vendors. This has had the positive result of creating a base of standards that guide product development but the negative result of taking time and sometimes producing inferior solutions in the interest of compromise.

Second, for all of the forward progress in openness, mid-tier systems from multiple vendors still cannot be integrated without considerable effort. While this might not be such a problem in the mainframe environment, where large procurements tend to encourage a single-vendor relationship, mid-tier systems are relatively inexpensive. That means that companies are more likely to buy mid-tier systems from a range of vendors to try them out only to discover many years after the trial period that they are faced with integrating many different vendors' systems. Also, the relatively low cost of mid-tier systems has enabled departmental development shops to make their own purchasing decisions over the years. When now trying to reengineer business processes across old departmental lines, these multivendor differences are showing up.

Perhaps the most interesting challenge to the mid-tier range of systems is the potential for desktop systems and applications to grow to meet the demands that are currently satisfied by mid-tier systems. Service providers are watching developments in new operating systems, such as Microsoft NT, very closely.

There are pros and cons associated with moving new applications to the high-end desktop environment, particularly as it has been dominated by a single vendor. On the one hand, systems integration may be easier to achieve. On the other hand, there is always the fear of lock-in. As one service provider's IT director said when trying to explain why the company might abandon its search for a strong multivendor mid-tier systems environment and rely increasingly on desktop products, "Proprietary lock-in stinks, but can it be any worse than what we are dealing with in the existing mid-tier environment?"

What Does Integration Mean at the Systems Level?

From the service providers' perspective, at least, three main elements constitute a pathway to integration. These are applications portability, manageability, and systems interoperability.

Portability: Getting more from applications investment

Portability is the ability to run an application on multiple vendors' platforms without change. Increasingly, this concept is being extended to apply to distributed applications, where pieces of the application can be run wherever it makes sense. There are several business advantages to achieving applications portability:

- The ability to distribute applications means that employees can access systems from wherever they are. A sales tracking application, for example, could be run at each branch office to track individual performance and could also be used at the corporate level to track results by region or across the company. Regardless of what computing environment exists at each of those locations, the application can be run if there is portability between the domains and between different vendors' platforms. Requiring internal departments to buy the same equipment and platforms won't eliminate this problem (even if it were possible), because service providers are likely to find the need to interact with alliance partners that use different environments. It will become increasingly important to achieve portability going forward.
- Service providers have traditionally built their own custom applications, including everything from billing systems to loop maintenance systems to forecasting systems. Some might argue that service providers have never been particularly skilled at software development, nor have they been efficient. They have had time on their side and deep pockets of money to spend. Not so anymore. In order to move faster and smarter, service providers are having to decide which systems really need to be developed internally and which might be better developed by companies that specialize in that field. There is an increasing trend on the part of service providers to buy whole systems from suppliers rather than buying pieces that need to be assembled and so benefit from commodity prices. But in order to take advantage of this buying strategy, while still being able to flexibly link the applications from different suppliers' systems, defined standard interfaces are crucial. Portability makes it possible for service providers to make use of off-the-shelf third-party software.
- There are cases in which service providers just might know how to build a better mouse trap and might want to export that skill to other service providers. If one company, for example, were to develop a gateway to be used to exchange trouble tickets, it might easily interest others in the code, both as a way to make it easier for the companies to achieve process integration and as a way to recoup development costs. Portability makes it possible for this to occur in a cost-effective way.

Manageability: Staying in control

Systems management across multiple computing domains is a mostly manual, and very costly, task. The parallels with network and service management are many, although many of the processes used to manage systems are quite different from those used to manage services and networks. Problems in systems management are not unique to communications service providers—they are shared by any company that makes extensive use of different types of computing systems. A few of these systems management problems are identified here:

Software administration. In a distributed environment, in which multiple users access the same applications, it is very important to be able to distribute, install, activate, and test both systems and applications software. As program bugs are found, for example, or new features or capabilities are added, the multiple copies of the software need to be updated to the new release without requiring a site visit to each computer.

User administration. In a company whose employees number several hundred thousand, keeping track of which users can access what applications and data and what types of communications capabilities are needed to access those applications can be a full-time job for a small army. As employees move from one job to another, they need to access some new systems and no longer need others. As whole offices move from one location to another, it is necessary to duplicate the existing computing arrangements before the group moves so as not to lose any productivity. Doing that without management tools is time-consuming and error-prone.

Systems configuration. Just as services and networks need to be configured to meet specific business needs, so do the systems used to manage or support those networks and services. The physical and logical configuration of systems and their components to provide the right capacity levels and functional capabilities is extremely difficult to do manually, particularly in a distributed computing environment. This involves keeping an accurate inventory of systems elements, their locations, and their interconnecting relationships and having the ability to change those relationships in appropriate ways to meet the business need.

Operations management. Again, the parallels with network management make it easy to understand what is required here. Computer centers, responsible for actually running the many applications and keeping the systems up, need to monitor, evaluate, and control workloads and operational states of the various systems.

Problem management. When problems occur, systems managers need to be able to detect the problem, analyze its cause, and direct recovery and resolution activities. This management need, similar to its network counterpart, requires the ability to monitor and log alarms and start or stop the reporting of alarms, either when a preset threshold is exceeded (such as when a major problem generates excessive alarms) or on demand.

Security management. Here, systems managers are concerned about preventing unauthorized access and detecting any violations of the system. Security management involves the use of authentication functions as well as security alarms and audit trails.

These are examples of the types of management capabilities needed to support systems management. Although individual vendors' products often provide some level of management capability in these areas, the real problem for systems managers is in trying to manage across vendors' products or across computing domains. When systems-level integration is the goal, the capability to manage seamlessly across environments is paramount to healthy and secure operation.

Systems interoperability: Creating links to facilitate information exchange

Interoperability is one of those terms that means different things to different people. In a systems environment, interoperability refers to the ability of any two systems to exchange data, whereas people worried about achieving end-to-end automated process flow-through use the term to mean not only the ability to communicate but the ability for applications to perform specific functions in response to requests.

The ability for any two systems to communicate requires that they be connected in some way, that they are capable of sending and receiving data to and from each other, that they are able to maintain a communications session together, and that they are able to "speak" a common language. In a proprietary environment, the challenge involved in making different systems interoperate is usually accomplished using primitive software that presents functional or performance limitations. Achieving systems interoperability in this environment requires internal development of code that will eventually be thrown away as service providers migrate to an environment that employs industry standards.

But even the standards that are now being employed do not completely correct this problem, since multiple standards exist for different purposes. Four important areas of systems interoperability standards—Internet, OSI, DCE, and SQL—are briefly introduced here to give an idea of the choices facing IT professionals as they gradually move from handcrafted links between proprietary systems to standard interfaces between "open" systems. It should be kept in mind that these are the standards that have the most bearing at the basic computing level of systems integration, as opposed to integration among service or network management applications (where, for example, the TMN standards would apply more directly).

The first two of these, the Internet and OSI models, are similar in that they are techniques for exchanging data between systems, and they address everything from physical connection right through to application services. The second two, DCE and SQL, address specific aspects of system-to-system interoperability.

OSI model. The open systems interconnection (OSI) model spans the entire range of system-to-system communications needs over seven layers. The lower four layers address the physical path between two systems and the ability to actually transport bits of data from one system to another successfully. Layers 5 and 6 describe the way information is presented and the ability to control a communications session. Layer 7, the applications layer, defines common protocols and services such as management, file transfer, mail handling, distributed transaction processing, remote procedure call, and so on. These protocols and services are used by applications to process requests and perform certain common functions.

Open systems interconnection provides a rich set of capabilities, which is desirable for many applications. However, it also represents fairly significant systems overhead and thus has not been as widely adopted as the Internet model. OSI application services and their related protocols are generally used when an application involves large numbers of transactions, where a secure environment is required, or where there is a need to positively track each transaction.

The OSI application services and upper-layer protocols (such as the common management information protocol, or CMIP) can be run over either OSI-based lower-layer implementations (such as X.25) or non-OSI lower-layer protocols (including TCP/IP), making it possible to interconnect over a wider range of data networks using these techniques. This is important because many service providers have existing data networks (many of them proprietary) that connect their computing sites. Without changing these data networks, it is still possible to write new applications using the OSI upper-layer protocols and services to communicate with other applications designed for an OSI environment.

Internet model. The Internet model is now widely used to support electronic mail and other data communications applications, with perhaps less overhead than that specified in the OSI model. Services include management, mail transfer, file transfer, and others. The Internet model is appropriate in situations where systems do not require end-to-end control of a transaction but rather are able to function on the basis of resending information when it is lost.

Similar to the OSI model, the Internet model contains both lower layers (the transaction control protocol and Internet protocol, or TCP/IP) and upper layers (such as the simple network management protocol, or SNMP). Generally, upper-layer Internet applications services are run over the Internet protocol (IP), so that applications that are written to run in an Internet environment communicate with other applications over a data network that is TCP/IP.

Distributed computing environment (DCE) and SQL (structured query language). DCE addresses systems services that are typically above the transport layer as it is seen in the OSI or Internet models. DCE includes a set of applications services and applications programming interfaces to support applications portability and systems interoperability in a distributed environment. The applications services include time service, directory service, and remote procedure call and can be run independently of transport. DCE is becoming a widely supported, vendor-neutral approach to intersystem communication.

SQL is an important common database access methodology supported by all of the key database vendors. Its importance is significant to interoperability (and hence integration) since it allows applications and systems to be written independently of the underlying database and data structures.

Why systems interoperability is important. It is probably obvious why interoperability is considered crucial in achieving systems integration. Without the ability to communicate in a common way across vendors' products and across domains, service providers are left to customize each interface when process agreements are reached. This is time-consuming, expensive to maintain, and fraught with problems as changes to either system may cause the interface to malfunction.

From a practical standpoint, especially where systems are distributed or where multiple systems of different service providers need to work together, the ability to interoperate is paramount. That is not to say that only one solution can be used; indeed, each of the four mentioned here will likely come into play to support different types of applications, since each fits a different set of functional, performance, and overhead characteristics. However, it is important that all suppliers implement the same version of the selected communications protocols so that interoperability can occur.

Interoperability is needed whenever two systems must exchange any type of data, and as mentioned, there are many applications services provided for this purpose. The ability to transfer a file or exchange mail requires the use of terms that are relevant to those applications, just as the exchange of management information requires a specific language. In the next chapter, we explore what it means to achieve interoperability of management applications.

Chapter

10

What Interoperability Means between Management Systems

Jim Herman, vice president of Northeast Consulting Resources, once commented that achieving interoperability between management systems is unlike most other undertakings. He observed that when two e-mail systems worry about interoperating, they are only concerned that the messages get to the right place intact. Management systems interoperability, on the other hand, is a bit like expecting those two e-mail systems not only to exchange messages but to open them, read them, understand them, and respond!

The term *interoperability*, in its broadest sense, captures this notion of two systems that are capable not only of sending data back and forth but of acting on that data to do something useful. Achieving interoperability means that process integration and systems integration have been achieved sufficiently to support the business agreement. And achieving interoperability cannot be done without standards, as shown in Figure 10.1.

Ideally, new standards should be defined only when a specific business need exists, i.e., top-down as in Figure 10.1. Business requirements identify where interoperability is needed, which in turn drives standards selection or creation. Implementation is more likely to be bottom-up, as companies establish a common interoperability infrastructure first and then make use of it to support specific information agreements.

In the last chapter, we looked at the communications models that exist for system-to-system communications interoperability and mentioned some of the applications services that have been defined to make use of these protocols. Here, we focus on the use of standards by applications to exchange service, network, or element management information.

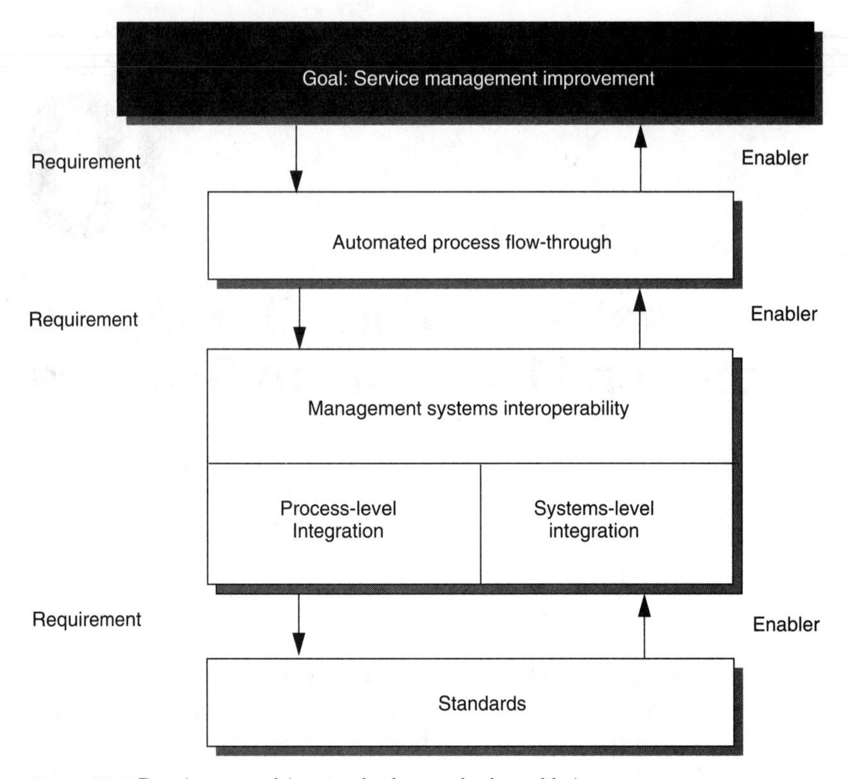

Figure 10.1 Requirements drive standards, standards enable improvement.

A Look at Standards

As a starting point, understanding a little about the standards-setting process is helpful. In the management systems environment, it is necessary to be familiar with international, regional, and de facto standards.

International standards

International standards bodies include the International Telecommunications Union (ITU), which we have already mentioned, and ISO (International Standards Organization). Historically, the ITU has been the place where the telecommunications industry convenes to set telephony standards, and ISO has been the place where computing standards are set. However, the lines between these organizations are no longer clear, since telecom companies participate in ISO and computing companies are involved in ITU. Fortunately, these two organizations have worked together to define management system standards, with both accepting the open systems interconnection (OSI) model as the base for management systems communication.

Regional standards

Regional telecom-based groups, including the European Telecommunications Standards Institute, Committee T1 (a group within the U.S. Alliance for Telecommunications Industry Solutions, or ATIS) and Japan's Telecommunications Technology Committee, become involved in management standards both as feeds to the international groups and as places where regional needs can be accommodated through extension to the international standards. Although these regional groups once acted quite independently, making it difficult for equipment and computing suppliers to know which standards to meet for their worldwide service provider customer base, they have begun to cooperate more closely with each other and with the international groups.

There are also regional groups established on the computing side of management-system standards. The U.S.-based Open Systems Environment Implementors Workshop, the European Workshop on Open Systems, and the Asia-Oceania Workshop share a common focus, which is to add implementation detail to the standards produced by ISO. These groups are collectively referred to as regional implementors' workshops. The network management working groups within these three organizations now work very closely together, dividing up the work to be done and endorsing each others' specifications. These groups add value to the process and help vendors and service providers achieve worldwide consistency. From our perspective, these three groups have done a great deal to move the industry toward a global approach to management interoperability. We applaud their efforts and encourage other regional groups to emulate their style.

De facto standards

Most de facto standards have grown up within the computing industry, which should be no surprise. From its beginnings, the telecom industry has set standards first, then built products. In the computing world, companies build competing products, the marketplace decides which is best, and standards codify the winning approach. A good example of this is the standardization of the distributed computing environment (DCE), first in the Open Software Foundation and later adopted by X/Open. DCE had its origins in two separate implementations—remote procedure call and directory—contributed by Hewlett-Packard and Digital, respectively.

Many different types of de facto standards are important to management system interoperability. The Internet Engineering Task Force has developed the simple network management protocol (SNMP) and related management information bases that define both the communications methods and the information to be exchanged among systems that are connected using the transmission control protocol/Internet protocol (TCP/IP).

Systems management standards, from groups such as the Institute of Electrical and Electronics Engineers (IEEE) and X/Open, address how computing systems are to be managed and how they will communicate with each other. Standards for object-oriented technology are emerging from groups such as the Object Management Group (OMG), which will have an impact on the development of next-generation management systems.

Putting Standards to Work

The mere existence of standards does not mean that the problems of service providers and their suppliers are solved. There are two serious problems involving management standards today: There aren't enough, and there are too many.

A bit paradoxical, but true. There are more standards published every year than anyone knows what to do with, since standards organizations sometimes seem to have no better measure of success than weight. The problem is that these standards often do not relate to a specific business purpose and so are either not applied at all or are inadequate. On the other side of the equation, where there are specific business needs to interoperate, the standards that exist often don't go far enough, and considerable work is required to turn a "base standard" into a detailed implementation agreement. That is where implementation groups such as the regional implementors' workshops, the Network Management Forum, the Electronic Commerce Implementation Committee of ATIS, and Japan's Interoperability Technology Association for Information Processing become involved.

These implementation groups apply the standards to specific business problems, making distinctions between the levels of specificity that are called for to solve different problems. Achieving systems-level integration, as discussed in the last chapter, involves applying sets of standards in a specific way so that all systems will be able to communicate, be managed, and so on, using a common base of techniques. Achieving process-level integration, on the other hand, requires not only a supportive underlying systems structure, but the use of combinations of standards and the extension of those standards to create specific (new) data definitions that capture the process-level agreements precisely.

Getting to the point at which end-to-end process flow-through can happen across a wide number of systems is not easy, but the technology to implement an underlying management systems infrastructure is now beginning to exist. A "critical mass" of supplier products is now becoming available, and we are witnessing a considerable upturn in the speed with which systems and applications are being introduced into the marketplace.

It's more than bits and bytes

To understand the issues around making management applications interoperate, understanding some of the basic standards is necessary. In the last chapter, we mentioned that the open systems interconnection (OSI) model is becoming important to service providers as a way of achieving systems-level integration. Layer 7 of this model is called the applications layer, and this is where all of the work specific to interoperability of network and service management applications is concentrated.

Within the applications layer, a number of issues must be resolved before management systems can interoperate in any meaningful way. A common communications capability needs to be implemented in both systems, and common services need to be employed that establish specific commands and terms to be used when one system asks another for information. Then definitions are needed to describe the types of operations to be performed. In an object-oriented approach such as OSI management, these definitions are called management support objects. Examples of such management support objects are a sieve, used to selectively filter the information passing between one system and another, and a log, which can be used to record transactions or events passing over an interoperable interface.

When two management systems have implemented common communications, common management services, and common management support objects, they have all the basic ability to exchange management information, but they still can't do anything useful. To make use of the capability requires yet more work, and here the "real" management applications come into play.

For example, if a service-level management system needs to reconfigure several service parameters, it might need to communicate with the network and element management systems that support the equipment or circuits used by the service. The service management system would send very specific instructions, such as "configure subnetwork A in the following way on Tuesday afternoon at 1400 hours GMT." It expects an answer from the network management system to say that the request has been received, the work has been done, and the system is standing by to activate the change at the appointed hour. It makes use of the protocol features to synchronize its request across multiple network management systems so that the service change can happen correctly.

As can be seen in this example, the ability to exchange information and respond to a common management language only sets the stage. To achieve the application described here requires that all systems have the same understanding of what is meant by "subnetwork A" and the configuration that is described, which brings us to a discussion of another type of objects.

Small Objects of Desire

Most service providers would probably agree that the widespread use of object-oriented technology for mission-critical, large-scale service management and network management applications is still a little way off. But that is not stopping service providers from gaining the benefits of object orientation in defining the information that is to be exchanged between systems.

Earlier we mentioned one type of object used in management systems, called a management support object. The other type of object used in management systems is called a managed object. Whereas a management support object (we used the examples of sieve and log) describes a management function that is reused for many different system-to-system exchanges, managed objects pertain to the physical and logical things that are being managed, such as equipment, a circuit, or a service. Once these managed objects are defined the first time, they can be reused in other applications, with different identifications applied simply by filling in data fields. For example, the definition of the managed object *circuit* contains attributes such as location and end point. By filling in the appropriate data, both systems are able to understand exactly what is to be configured.

Managed objects can and frequently do exist in the database of systems that are not object-oriented. However, a system that uses object-oriented design and implementation for the management application and uses object-oriented techniques for modeling the behavior of the networks and systems that are being managed is simpler to construct and test. With the commercial availability of object-oriented language, design tools, and databases becoming ever more common, this seems to be the trend for the future.

Getting close is getting nowhere

Procuring against standards can give a false sense of security, because when standards aren't applied correctly, or when they are applied with insufficient precision, the result might not meet the business need. Such is the sophistication of management systems that the level of handshake required between them is difficult even when both systems are owned and operated by the same company. Reaching effective agreements across company boundaries and involving multiple systems is far more difficult and requires a level of specificity that is unmatched by any other type of interoperable application. Thinking back to the farming and fishing communities of an earlier chapter, basic management system interoperability is akin to establishing a postal service and addressing scheme to exchange messages, as well as the paper and envelopes used to write letters and the language to be used. But having these communications facilities does not mean that if one person writes "Please sell me one," the recipient will know that the originator really means to buy a two-pound salmon on the next market day. In systems terms, getting this specific means establishing detailed implementation agreements between applications.

When service providers try to procure management platforms or management systems, they often don't apply enough specificity to their requests. It is not at all uncommon for service providers to request "TMN-compliant" systems, because they have heard about TMN and see it as a "good" or "safe" way to build systems. Unfortunately, if their request is not specific—if they just ask for "TMN compliance"—they are likely to get one of two responses from suppliers:

- "What exactly do you mean? If you tell me which TMN standards you need I'm sure we can meet your requirements, for a price."
- "I'll be glad to."

The first answer is at least honest, although service providers will be frustrated to hear it. The fact is that there is no way to measure "TMN compliance" except by defining exactly which of the ITU recommendations are to be implemented, how they should be implemented, and what information agreements must be capable of being supported. Asking for "TMN compliance" is like asking to "please sell me one," and honest suppliers are just as frustrated as service providers that they are unable to respond to such a general request. Those with fewer scruples (or, to give them the benefit of the doubt, a misguided belief that they know what TMN compliance is) will be happy to give service providers what they've got, which may in fact comply with some aspects of TMN but which will probably be useless if the objective of the service provider is management systems interoperability.

Given the complexities involved in achieving system-to-system interoperability, it may be obvious that the sooner the service provider can start working with the suppliers, the better. To the extent that service providers can include their suppliers in the detailed requirements analysis phase, the design phase, and the specification and development phases, interoperability between a service provider's service management systems and its suppliers' network and element management systems will be much easier to achieve.

In the same way, working with other service providers to decide what business needs must be satisfied by exchanging information in an automated way—from detailed requirements work right through to development—will yield a better result. Unfortunately, mere agreement to implement basic standards does not bring direct benefits but simply sets the stage. Use of these standards requires that they be applied to specific business needs. A management interface to exchange trouble ticket status information will be completely different from one in which the purpose is to configure or perform surveillance on network elements. Constructing agreements for each management need can take an incredible amount of time, and this is bad news for service providers, which don't want to wait for yet more "standards" to be developed, and which don't want to have to pay for customized solutions to every one of their management problems.

Although agreements need to be precisely tailored, that doesn't mean that each implementation agreement starts with a clean slate. Reusability is very important to speed the process and keep costs down. Unless agreements are constructed so that pieces can be reused, the time to develop interface agreements can be prohibitive. Building agreements with reusability in mind considerably reduces the cost of subsequent development agreements. An object-oriented approach helps with this problem.

Object orientation is very important to reducing the complexity of management system interoperability. Object definitions serve as reusable building blocks that can be combined in many different ways to describe different physical or logical resources. This building-block approach lends itself very well to data abstraction, which is one of the key requirements of process integration. In the next chapter, a framework for achieving this data abstraction using object-oriented technology is reviewed.

Chapter

11

An Integration Architecture for Service Management

The demand is rising for systems to interoperate in a diverse, multivendor environment to deliver increasing levels of automated process flow-through. However, no single technology provides all the answers, so different technologies must be used in combination to provide solutions that meet today's service management needs. In addition, there is a desire to take advantage of new computing capabilities, specifically object-oriented technology, as they become more widely available. In order to successfully chart a path from today's systems environment to emerging object-oriented technologies, a target architecture and evolution path are required.

To understand the architecture, it is necessary to bring together many of the concepts discussed earlier. These include horizontal and vertical integration, abstraction, partitioning, and recursion, which are complex taken individually and very difficult to deal with taken together. A group of very bright people working together within the Network Management Forum have developed a technical strategy to show how these objectives might be achieved, first with existing technologies and moving over time to a more encompassing object-oriented approach. We have used their document, the OMNI*Point* Integration Architecture[1] as the basis for this chapter.

---

[1]Network Management Forum. 1995. OMNI*Point* Integration Architecture. TR 114. Developed by Stuart Haines (GPT), Norman Kincl (HP), John McKenna (IBM), Chip Kerr (EDS), and Shekar Sundaramurthy (AT&T).

An Architectural View of Function and Data Abstraction

As part of the earlier discussion of service provider requirements, we covered the business need for abstraction from one level of management system to the next. From an architectural perspective, there is a need to consider abstraction at both the functional level and the data level. *Functions* refer to management operations (e.g., event handling, problem management, ordering, planning, and design, etc.), and *data* refers to management information (e.g., user, service features, location, type of equipment, etc.)

In Figure 11.1, the horizontal axis shows that the more types of information a system must deal with, the greater the need for abstraction—to mask the intricate details of data format and allow a common way of handling information across systems. In the same way, as shown on the vertical axis, the more management functions that must be performed by a system, the greater the need for abstraction in the management functions—again, to take out unnecessary details.

The first quadrant (lower left) represents basic element management. Many existing element management systems provide solutions in this area, and their management functions have a direct relationship to the opera-

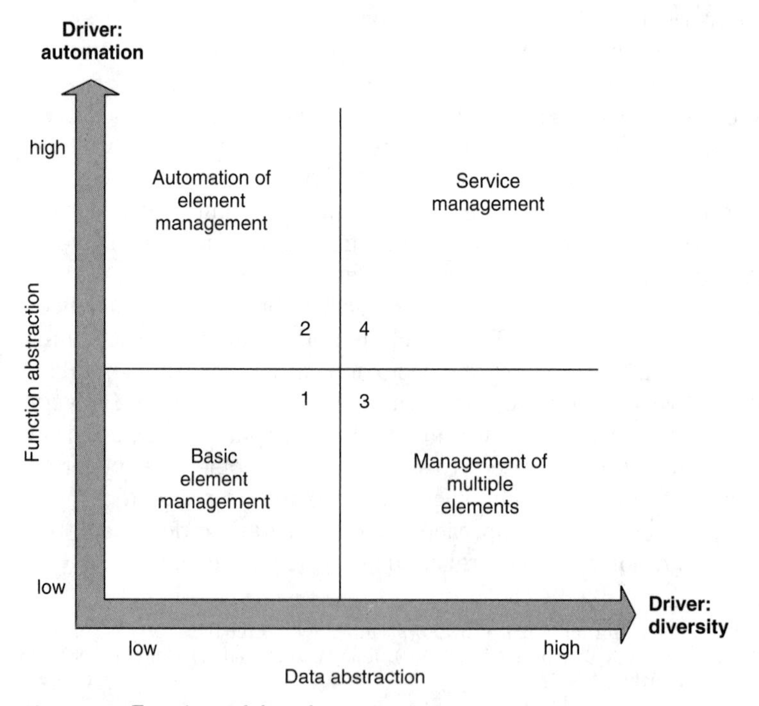

Figure 11.1 Function and data abstraction.

tions of managed elements such as switches, transmission systems, and so on. These functions or capabilities are tailored for the individual elements and hence cannot easily be used by other systems, especially those requiring a higher-level view of the information. For example, one multiplexer management system might have a function called "report alarm" while another might handle a very similar set of circumstances with a function called "increment counter." A network manager trying to interpret what is happening across these two systems simply has to learn the two different ways to access information—requesting the first system to display alarms while requesting the second system to show the contents of its counter. It can be very difficult (if not impossible) to automate the link between the network management systems and these element management systems, because the functions are not common.

Differences in the data models of the managed elements will also exist in systems found in this quadrant. For example, one system might represent a serious problem as a "severity 1 fault" while another may call it a "class 5 alarm." When each management system uses its own data representations, almost no data abstraction is possible. When all management systems occupy this lower left quadrant, the only way to deal with the resulting complexity is with people and their skills and experience.

The movement towards becoming a lean service provider will drive a provider into different quadrants of Figure 11.1. For example, the drive to simplify and take cost out of management processes drives automation of repetitive tasks. This movement is represented by quadrant 2 (upper left). Some management systems have evolved in this direction, either through simple conversion from one format to another or using more complex data-modeling techniques. Beyond simple functions such as event monitoring, however, there is a limit to what can be achieved in this quadrant because the underlying data models are still different, and nothing has been done to insulate these differences from the management functions.

By encapsulating the data representation of each managed element, a single generic application can manage many dissimilar elements. This is a move toward the third quadrant (lower right). It requires that each system supports a common data format, usually by performing a translation, either within each system or externally. If data formats aren't common, the management application must make a change whenever a new type of managed element appears. And for each management function, agreements on common data formats must be negotiated between management systems.

Traditionally, the move along the two axes has been accomplished by individual service providers without coordination and with considerable expense. Fortunately, advances in object-oriented technology and information-management technology make it possible to coordinate an approach to function and data abstraction.

These advances make it possible to move to the fourth quadrant (upper right) and manage effectively and efficiently at the service level. The management systems framework, discussed next, is focused on the requirements of this quadrant.

Management Systems Framework (MSF) Overview

To fully address the many technical challenges of service management, the management systems framework integrates an object-oriented framework with a manager/agent framework. Let's look at these two briefly.

Object-oriented framework

The object-oriented framework enables application interoperability and distribution using the principles of function and data abstraction that were just discussed. Put more simply, by reducing all functions and data to commonly defined "objects," it is possible for an application to use objects in different combinations to produce a specific result. For example, by combining the "report alarm" function with "severity 1" data (pretending for the moment that these are actual standards that have been implemented in element management systems), it would be possible to develop a network management system application to request that the element management systems alert it to critical faults.

Manager/agent framework

The manager/agent framework is based on the assumption that for any type of interaction between management systems, one system acts as the "manager" and the other as the "agent." The manager system requests information about something (a switch, for example), and the agent system provides it.

Management systems can either assume the manager role, the agent role, or both, although for any one interaction between systems, an application must be seen as one or the other. For example, a network monitoring application may act in the manager role when requesting alarm information from "agent" element management systems, but it may function as an agent when asked by a service management application to provide overall status of the network.

Using this framework (and the standards such as TMN that support it), the only way to achieve interoperability between two systems is if one has implemented the manager side of an implementation agreement and the other has implemented the agent side. Two "manager-only" applications cannot communicate, nor can two "agent-only" systems.

Communications between systems can be through a management communications protocol (such as CMIP or SNMP) or using a client/server computing

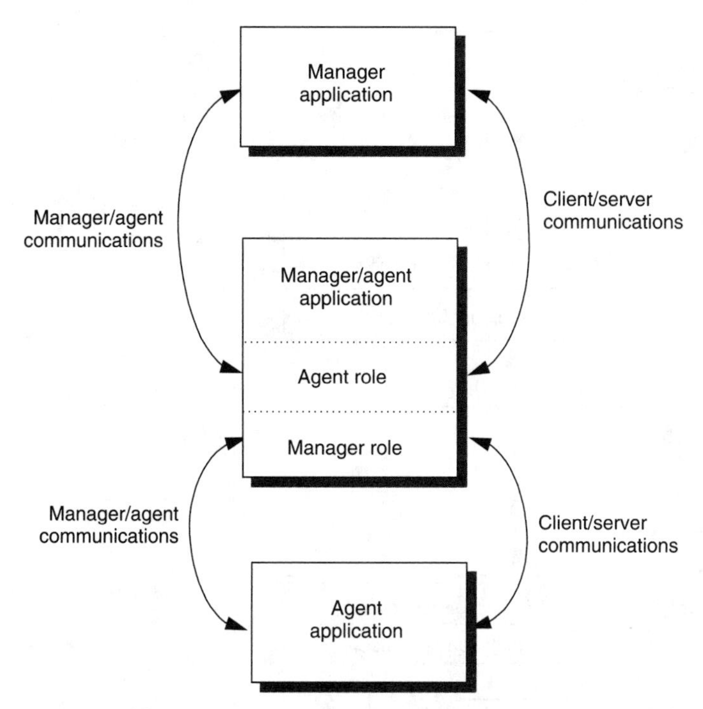

Figure 11.2 Management systems operating in a hierarchy.

mechanism. Figure 11.2 shows these options, as well as the hierarchical manager/agent relationships that might exist among multiple management systems.

Using client/server, the "manager" application would reside on the client system and the "agent" application would reside on the server. The two applications would use regular computing communications techniques to exchange information.

The manager/agent framework is based on the use of computer representations of real resources, called objects, which we introduced in the previous chapter. By representing data as objects, it is possible to set the stage for the use of an object-oriented framework to fully exploit the advantages of an object-based structure. In essence, this can mean that everything can be seen as an object, including individual resources (e.g., circuit end point), groups of resources (e.g., circuit path), or even entire applications (e.g., report a critical alarm on a circuit path). Using higher-level object-oriented techniques, applications can be considered "objects" in their own right and can be used recursively in the creation of new applications.

Figure 11.3 shows the full management systems framework. In a moment we will review the elements of the framework, but first let's take a look at its general characteristics.

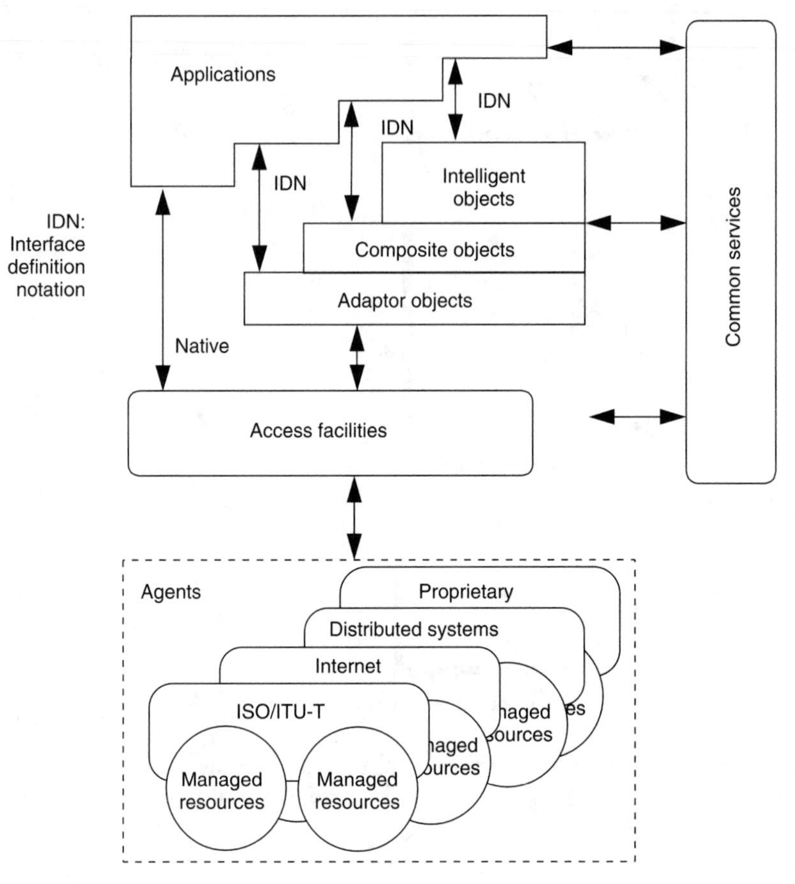

Figure 11.3 The management systems framework. *(Network Management Forum. 1995. OMNIPoint Integration Architecture.)*

At the very top of the diagram are the applications and at the bottom are managed resources, which could be switches, circuit paths, or whole networks. In between are represented several different techniques that could be implemented in middleware to support applications that will ultimately control these resources. Each of these techniques (such as the direct use of access facilities or the use of composite objects) represents a different level of sophistication and requires different types of underlying computing technology, some of which is not yet generally available.

Let's consider an example. A service provider wishes to streamline the processes used to provision its private-line services. Part of the application at the network level might be to configure a circuit by manipulating objects such as "endpoint" and "channel" to make changes in the underlying physical network using simple access facilities. This network-level application could, in turn, be represented as a "composite object" invoked by a service-level application along with other configuration management application "objects" to permit individual customers to modify the capacity of a communications path as demands change. Another, higher-level service management application might be used to handle all of the functions necessary to provide service to a new customer. Here, a "change management application" object might be invoked along with "service level agreement application" objects, "trouble reporting application" objects and "billing account application" objects to establish all of the necessary records and direct all of the work activities that are necessary to provide service.

These increasingly abstracted and complex objects are defined in the management systems framework as a hierarchy. Think of them as similar to Lego™ kits that contain everything from simple bricks to complete ready-built modules.

The Components of the Management Systems Framework

The management systems framework consists of the following components:

Agents

Keep in mind that any interaction that occurs between two management systems requires that one system act as the "manager" and the other as "agent." In the management systems framework, agents represent the link between the resource and the management system. For example, a line card might be capable of having its ports turned off and on. An agent for that line card would provide names for those ports (managed object definitions) to make it possible for a management system to request that the configuration of the ports be changed. Managed objects provide the management view of the resources. The managed object definitions identify the set of functions and information relevant to the network or service element being managed.

Access facilities

Access facilities support communication between applications and managed resources (via the agent) through protocols such as the common management information protocol (CMIP), the simple network management protocol (SNMP), remote procedure call (RPC), or a proprietary protocol such as IBM's SNA. Applications that rely only on these access facilities for their communication with the managed resources will use native interfaces that match the specific environment for which the object definitions are written in the agent. For example, in an open systems interconnection environment, objects are defined using the guidelines for the definition of managed objects (GDMO), while in an Internet environment, objects are defined using the structure of management information template. Because each of these environments is highly structured, applications that are built using only access facilities and native protocols can achieve only minimal abstraction of functions or data.

Adaptor objects

Adaptor objects eliminate the need for applications to manipulate the individual management protocols of the access facilities. In other words, the use of adaptor objects makes it possible to begin to even out the differences between native communications methods, giving the applications developer a common way to address each resource regardless of how it might be described in its native form. Although there are still differences among the functions that can be performed on each resource (because of native protocol limitations) at least at the data-access level much of the complexity is removed through the use of adaptor objects.

Composite objects

Composite objects take function abstraction and data abstraction a step further. Whereas adaptor objects simplify the way an application accesses data or requests an operation to be performed for single resources and single operations, composite objects do the same for groups of objects and basic sequences of operations. Think of composite objects as offering an extended object model with the following characteristics:

- Object groupings are simple. Objects have mostly static relationships, which are encapsulated in the form of another, higher-level object.
- Operational sequences are well known. They are typically predefined. Should a failure occur, an algorithm defines the recovery sequence (an application doesn't have to make choices about how to recover).
- Generic functions are possible. They may involve operations that can be applied in the same or similar way across some or all of the objects in the group.

Intelligent objects

Intelligent objects take function and data abstraction to the highest level. They allow applications to interact with objects representing elements of the service and make it possible to synchronize the many different events and operations that might affect a single resource or a group of resources. Compared with composite objects, intelligent objects offer significantly enhanced characteristics:

- Object groupings can be complex and dynamic. While servicing an application's request, relationships among affected objects may change dramatically and new objects may be created. An application program or user may be unaware of many of these underlying objects and their interactions.

- Intelligent objects can learn operational sequences. Depending on the results of an operation, rules-based expert systems can control their behavior. The objects may dynamically contribute additional rules during their operation.
- Functions become service-oriented and deliver complex operations across a diverse collection of systems and processes.

Common services

Common services provide support to the other components of the management systems framework. Some examples are directory services, object-class information services, and security services. These services have standardized interfaces. Applications may use these in different ways and for different purposes.

Applications

Applications perform various management tasks such as order handling, problem management, and so on. The applications typically reflect the policy and disciplines that fit a particular service provider's environment and will range from low to high complexity. Recalling the TMN model, applications can be developed to manage elements, networks, or services. Using composite and intelligent objects, a sophisticated service management application might well contain objects that represent network or element management applications.

Evolving toward the Management Systems Framework

The components of the management systems framework offer progressively more abstract (and comprehensive) services. The move to higher-level capabilities, through the use of adaptor, composite, and intelligent objects, will evolve over time as the supporting technology becomes more readily available. In effect, this is a little like hi-fi systems 20 years ago. Once, you had to have intimate knowledge of the specifications of the amplifier, deck, and loudspeakers to make the system work. Today, you buy prebuilt modules that plug together—abstractions from the details. As suppliers gain experience with these technologies, higher-level objects will appear, and this will bring key benefits:

- For any specific management function, more work will be performed through the capabilities of the management platform than by the application. This means applications become simpler and quicker to develop.

- Functions and data are available to applications at a higher level of abstraction that is more useful to a service management view than a network management or element management view.
- Applications can have a fairly simple structure yet be able to perform highly sophisticated tasks.

Achieving the management systems framework is an evolutionary process. By addressing the most basic needs first and then building on those beginnings, it is possible to gain measurable value at each phase of development, while ensuring that all needs can ultimately be met.

Developers are only now beginning to implement the basic manager/agent framework contained within the management systems framework. Objects are being defined to represent managed resources, and objects are being grouped to achieve specific management objectives using the common services and access facilities currently available.

Moving towards the full management systems framework will take time. First the basic technology must be fully developed and stabilized, then platform developers must implement the capabilities, and, finally, applications developers must become comfortable using the technology. Still, by articulating a technical vision, it is possible to see how today's technologies must be viewed—not as self-contained all-purpose solutions but as parts of a more comprehensive and sophisticated framework.

Chapter

12

Barriers to Service Management Excellence

Let's sum up the story so far. The communications industry is becoming ever more competitive, and the "triple whammy" of the three-legged stool—keeping profits healthy when prices are falling (by reducing costs and increasing revenues), continuously delighting customers (by improving customer service and quality), and building new revenue sources (through faster service introduction)—cannot be solved just by reorganization and downsizing. Neither can it be solved by new entrants that look like slimmer versions of the existing telcos—something that is likely if the new players are run by ex-telco people.

Reengineering and automating end-to-end process flows is key to sustainable service management excellence, but as we've said, achieving this goal requires a significant amount of effort by the service provider working in conjunction with its suppliers. No matter how hard a service provider might work to achieve success, the integrated nature of the global communications industry and the complex value chains that exist within it ensure that no one company can solve all the problems on its own.

In this chapter, we describe the areas where the industry as a whole—both operators and suppliers—needs to cooperate to solve the wider problems of end-to-end service management automation to support the service providers' quest for service management excellence.

Aspiring lean service providers face two types of barriers when attempting to achieve improved service management process flow-through:

- The computing infrastructure might be incapable of supporting a wide variety of service and network applications or increasing levels of integration. Platforms are often either of insufficient industrial strength or unavailable as commercial, off-the-shelf products.

- There is resistance to reaching common process interface agreements to support either the administrative information exchange with customers and other service providers or links to networking equipment.

These barriers are examined in turn.

Resistance to Systems-Level Integration

The established telecom service providers are among the largest corporations in the world, and their systems requirements dwarf those of most companies. An operator may have many millions of customers, millions of pieces of networking technology, and hundreds of thousands of network nodes, representing a huge data-storage challenge. In addition, the nature of the communications business involves real-time or near real-time transactions at a tremendous rate. While a typical large private-enterprise network might generate a dozen network events in a minute, the networks of telecom service providers generate hundreds or thousands.

Service providers have not always benefited from the move toward open, interoperable systems platforms. One reason for this is their size and performance requirements, but another is that monopolistic providers became accustomed to developing their own detailed requirements for large systems. One European service provider executive told us "Our development people never look at off-the-shelf solutions. They have a mindset that everything is unique to the telecom industry and must be handcrafted. They would build their own operating systems if we let them." Lean service providers can no longer afford this luxury. The cost and time involved in developing detailed requirements, often paying custom prices to computing manufacturers to deliver and then bearing the multiyear cost of software development and maintenance, is an expense that must be rapidly eliminated.

However, there are significant challenges to off-the-shelf solutions. Mid-tier computing platforms available today often don't meet the scaleability needs of service providers for mission-critical applications. Nor is it possible to tightly integrate multiple computing platforms—whether from the same vendor in the same domain (i.e., mainframe, mid-tier, or desktop domain), from different vendors within the same domain, or between domains.

Computing companies continue to balk at reaching agreements among themselves that will help buyers and users of their platforms to more easily achieve systems-level integration. They sometimes act as though by retaining their own proprietary methods of doing things they will keep customers. This is the old model, and except in instances (such as the current dominance of Microsoft or the once-dominant IBM) in which a single supplier has captured the market, it just doesn't wash.

Akihiro (Kei) Takagi, executive manager of NTT's multimedia business department, is very clear about his company's business reasons for needing a standards-based computing platform. But he makes the telling comment that suppliers often care more about competing with each other than about satisfying their customers, giving as proof examples of things supplier product managers (certainly not the salespeople) have actually said to him or to other large companies in response to stated requirements:

- "We don't believe this is a real user requirement."
- "Users shouldn't demand this kind of standardization."
- "We can provide a much better solution than any open solution."
- "We understand the users' needs, but we have our own business to run."[1]

If the service provider industry—without doubt the largest and most concentrated market for computing hardware and software—is treated with such obvious disdain by its suppliers, what chance do small- to medium-sized users have? The irony in all of this, as Takagi so rightly points out, is that suppliers are so busy watching their peers, they lose all sight of threats from outside their circle. Manufacturers of mid-tier computing products, which have yet to take the service providers' requirements to heart, could find that they might all lose the battle together, letting the desktop segment rise to dominance.

We would like to be charitable and say that suppliers don't mean to be obtuse—it's just that the right hand doesn't always know what the left hand is doing. And that is certainly true of the service providers, as well, which may still be giving suppliers mixed signals about what is needed.

Service providers can help themselves by placing a common set of requirements upon computing suppliers. This would help in two ways—first by reducing the amount of effort needed by any individual company in documenting requirements, and second by taking advantage of purchasing clout to encourage computing suppliers to meet their needs. Although it is understandable that computing vendors do not wish to be considered commodity suppliers, it is our belief that any vendor that can meet the service providers' needs for interoperability, manageability, and portability across computing domains will have access to a huge market.

A system is not a system . . .

Service providers often make a distinction between "general-purpose" systems and "network management" systems. Often, these system types will

[1]Takagi, Akihiro. 1995. Escaping Supplier Monopolies. From a presentation given at the ETIS '95 Telecommunications and Information Technology: Connecting Customers to Suppliers Conference, May 31–June 1, 1995, Sophia Antipolis, France.

be controlled by different groups of people: The IT professionals are generally involved with the more traditional general-computing systems, and network operations teams are usually involved in network-facing management systems. This distinction is becoming less and less relevant as computing migrates to more standardized distributed systems and as service providers break down their once-rigid departmental barriers.

If service providers do not take steps to integrate not only the computing infrastructure but the cultural groups that are responsible for them, then this distinction will begin to cause more and more difficulty as they try to link their service management processes. For one thing, service management by its definition involves both general-purpose systems to support such things as ordering and billing, and network management systems to support service configuration, performance monitoring, and so on. These need to be linked, which will be easier to accomplish if the underlying computing technology is common. For another thing, specifically tailored network or element management computing systems—especially those used to support real-time functions—are becoming too costly to maintain, and suppliers are increasingly trying to adapt general-purpose systems to network management use. If the two worlds are not linked, the computing suppliers are likely to get mixed signals, which could well delay progress toward fixing many of the problems of the systems environment.

One way to think of the different computing needs is that the general-purpose elements should be common across all systems, including those used for "back-office" functions as well as those used to configure a customer service. On top of this base capability, specific management applications services can be specified in order to support the breadth of service, network, and element management applications needed by service providers.

If this view of systems requirements were shared by IT as well as telephony people within a service provider company, they would be in a much stronger position to influence suppliers and to gain support for systems-level integration.

Resistance to Process Flow-Through Agreements

The barriers to systems-level integration are significant, but they are not the only hurdle service providers face when striving for service management excellence. At the process level, the problems are different but no less formidable.

In a world increasingly characterized by "coopetition"—simultaneously competing and cooperating—agreements with other providers, customers, and technology suppliers play a key part in unlocking end-to-end process flow-through. But at all points of interface—with suppliers of networking technology, with other service providers, and with customers—there are pockets of resistance.

Resistance from equipment suppliers: The fear of becoming a commodity

As the liberalization process gathers speed, more and more service providers are procuring network equipment from multiple suppliers. This is often a first reaction when ties to the nationally favored supplier are severed by legislation, and it continues as service providers strive to keep current with the latest technology. But all too soon, the pain and expense associated with managing multiple vendors' equipment across different and proprietary interfaces shows up. This is one of the first areas that the lean service provider needs to address—work to define common interfaces that allow multiple suppliers' equipment to be managed in a common way.

However, such agreements are dependent upon communications equipment suppliers' cooperation to support process flow-through agreements if full automation is to be achieved. Often, the response from equipment suppliers is much slower than service providers would like. The main reason for questioning the necessity for interoperable interfaces is the fear that their equipment will become a commodity if its management interfaces are open and common. In the view of the service providers, nothing could be further from the truth. Consider the following points:

- Network and element management systems that fit within a service provider's service management framework will have a competitive edge over those that do not. Indeed, as many service providers adopt automated process flow-through techniques to survive in their competitive marketplace, they simply will not be able to deploy a supplier's system if it cannot interoperate. They cannot afford the financial overhead associated with performing manual integration throughout the life cycle of the equipment. Nor can they afford the commercial costs associated with falling short of service excellence goals. In other words, they will not buy.
- The favored status many suppliers once received from their country's monopolies is waning rapidly as competitive service providers begin to buy on a worldwide, price-performance basis. New networking technologies, new supplier entries, and the reengineering of service management processes are signaling the time for change.
- Unique interfaces do not provide competitive differentiation. Features and price/performance provide that edge. A unique interface may actually inhibit equipment sales, whereas a common interface is likely to spur them. The first suppliers to respond proactively to the service providers' requirements for common interfaces stand to gain a significant share of a significant market.

However desirable it might be for service providers to gain the support of their suppliers to support common interfaces, some who are on the front line wear the scars of their experiences. They have found that general agreements often aren't enough to produce real benefit and that the only way to get results is by getting specific—sharing business objectives and operational process definitions with suppliers and jointly designing a solution to the problem. This can be extremely difficult to do, for two reasons. First, many service providers just don't have an overall view of how their processes work (or should work), and those that do may be very reluctant to impart such highly competitive information to any outside company. Second, service providers don't seem to have a good track record when working across departments or product groups, much less with totally separate companies. For these reasons, it is likely that service providers will enter into strategic relationships with a very limited number of computing and telecom equipment suppliers and concentrate their efforts with those companies instead of more generally.

Resistance to interfaces with other service providers: Consorting with the enemy

Most interprovider agreements, in the few places they exist, have been developed on a pair-wise basis, and most providers have been reluctant to subject alliance-based interfaces to industry agreement. Recognizing that process flow-through is key to competitive success, they once took the view that any agreements they might have reached needed to remain proprietary. This position is quickly becoming impractical as providers expand their reach to automate process flow-through with greater numbers of other providers and the sheer cost and complexity of interface technologies expands exponentially.

Even dominant carriers, which might once have tried to enforce their own methods on partners, are finding that old strategies aren't effective in the new environment. Said one representative of a major U.S. telecommunications company, "As we develop partnerships with other providers, we know there will be some issues that will be critical for us to 'win.' We don't yet know what those will be, but we're quite certain that process-interface agreements are NOT among them. It's far better for us to say 'Let's all adopt common industry agreements' than to push our own ideas in this area."

Beyond not wanting to seem like bullies, service providers will find good business reasons for reaching industry-wide agreement on interprovider interfaces. For one thing, the industry is changing so fast that service providers cannot be certain that today's partners will still be allies tomorrow. For another, although the lack of a common interface can inhibit full process flow-through, the mere fact of having a common interface doesn't rule out the ability of one provider to make better use of the common data and therefore gain competitive advantage. Those that are able to incorporate common industry data definitions in their internal processes will eliminate costly and error-prone translation and will benefit far more than those whose internal applications continue to require translation of data at each system-to-system interaction.

Although there are plenty of examples in which enlightened managers see the need to cooperate with "the enemy" in certain areas, there are as many or more instances of managers at all levels resisting this vehemently. One of us had an interesting discussion with an executive of a large global provider several months back. Commenting that we were pleased to see his company finally beginning to participate in industry activities to forge process-level agreements, his response was strong and immediate. He vowed that his company "would never work with" a certain competitor (which was also involved in these activities) and promised that he would take immediate action to pull his company out of any such collaborative work! This type of knee-jerk reaction is not uncommon from service providers that are unable to distinguish between areas where competition is essential and areas where the entire industry stands to benefit from collaboration.

Resistance to interfaces with customers: Giving away the store

Although service providers can see clear benefits to having their suppliers support open interfaces, when they are the supplier their thinking becomes more cloudy. The value of industry-standard interfaces with customers remains difficult for some service providers to understand. For whatever reason, including fear of losing account control, fear of being directly compared with competitors, or fear that investment in customer-site systems will be jeopardized, the benefits of a common customer interface are sometimes overlooked.

But from the viewpoint of the customer, the need is remarkably similar to the service providers' desire to establish common interfaces with their suppliers. Enterprise customers employ the services of more than one provider, and they certainly don't want to build multiple proprietary interfaces into their own support systems and processes. Some of the more advanced service providers that are pushing for agreements with customers see the need to minimize the number of separate, customer-tailored requests for automated order transmission, service level agreements, and performance reports. Where industry agreements are available, these service providers believe they can encourage their customers to make use of them, whereas it is much more difficult to encourage a customer to accept the service provider's own proprietary approach.

A growing number of service providers see more general benefit to achieving common agreements with customers. They recognize that while customers will see positive value of an automated interface with providers, they do not choose a provider based on the way the interface is constructed, except perhaps to negatively view providers that require something unique. Once a decision is made to automate an interface, the only interface of value is one that fits without requiring extra training or development to accommodate. Norman Meyers, a veteran of GTE's business process reengineering effort, said "Our customers will demand that our industry provide common ways for them to do business. Those of us who can be first to show we're on their side will make it clear we've got their interests at heart, and that can only help us as we expand into new services."

Even with standardized interfaces, there is enormous scope for service differentiation. Automobile manufacturers, computer companies, and many other industries have shown that standards (from safety regulations to technical interfaces) can be turned to significant proprietary advantage with a "one step beyond" or "standards plus" mentality.

New Services Could Make Integration More Complex

Many service providers and suppliers are now turning their planning horizons toward a broadband, multimedia "Infobahn" world. Surely in the 21st century, all of these problems will be swept away with new technologies and processes, right?

Well, the reality seems a little different. At the transport and switching levels, networking might get a bit simpler, provided the suppliers of systems such as ATM build in the necessary management capabilities and means of interfacing to other systems.

On the business front, however, life is set to become more complex. Certainly it is faster to introduce new services via stovepipes and overlays, and there is little evidence that service providers are moving away from this approach. Wendy Shorrock of BT said, "Fast-track solutions are increasingly the 'business requirement,' which makes it harder to absorb these solutions into the integrated whole later on."

In addition, the value chains that will deliver that home-shopping service or video-on-demand to your kids will be the result of several players acting together, as shown in Figure 12.1. This means that the interfaces across industry boundaries for processing orders, resolving problems, billing, and so on will involve organizations that today are not in the chain at all. Thus the complexity will only increase.

Developing seamless, end-to-end service management processes today, regardless of industry segmentation and the technology that might eventually be used to deliver the services, is a prudent course of action. To the extent that industry agreements are widely accepted, they can be used by the new players. However, in the absence of firm agreements, these new companies may inadvertently develop all-new solutions and may set integration objectives back several years.

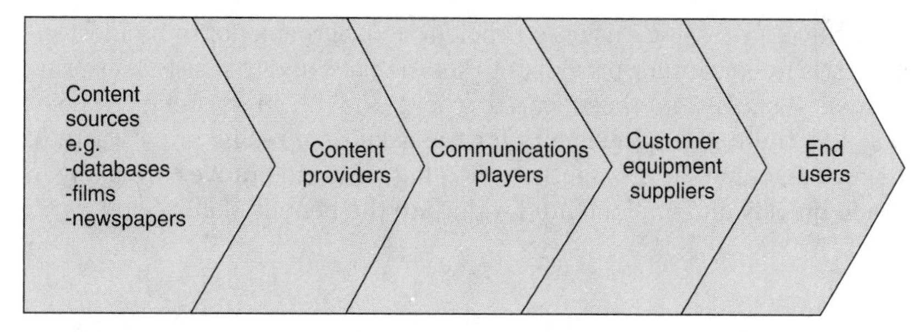

Figure 12.1 Information services value chain.

Time Is of the Essence

Although the telecommunications industry is increasingly under pressure to change, and even though there is a relatively small number of key "boundary" agreements that really impact service management excellence, industry agreements take time. Service providers cannot wait for industry decisions to begin streamlining their internal processes; the pressure to remove cost and improve service quality is simply too great. So they will begin to make decisions about how to link their management processes, even though these decisions may conflict with agreements ultimately reached by the industry. For example, a service provider will need to settle on a single representation of "customer name" across its ordering, installation, and billing systems in order to take advantage of a common customer database. The definition and format it chooses might not match those eventually selected by the wider industry in order to exchange orders, billing records, or trouble tickets.

Over time, service providers should be able to make use of common industry agreements (those that govern their external links with equipment suppliers, customers, and other suppliers) within their internal system-to-system handoffs. The move to replace one term with another can be made gradually, with translations filling the bill until a change can be made. So it is not necessary for the internal systems reengineering and external process flow-through agreements to be implemented at the same time, provided the service provider is cognizant of the need for translation.

Of more concern than eventually making the shift to take full advantage of industry agreements is that service providers might hold tight to technology they have already implemented internally, to the detriment of reaching any industry agreement at all. A challenge to service providers will be to refrain from protecting their internal investment decisions in the interest of making rapid progress toward industry agreements.

Service providers will do what they must to stay in business. They will not embark on industry standardization efforts that are not absolutely necessary to solve a pressing business problem. For agreements to be most effective, it is important for providers to set the pace and to resist sponsoring agreements except in those cases where the business benefits point strongly to the need. Agreements for the sake of agreeing are a waste of time. Agreements that are tightly tied to business imperatives are likely to be made quickly and implemented widely, to the benefit of all.

Part

3

". . . It's Best Not to Go It Alone"

So far we have painted a pretty dismal picture for service providers that don't see the need to make serious changes to the way they do business. We stressed the importance of balancing the three-legged stool by simultaneously improving service, reducing cost, and shortening product introductions. We postulated that only "lean service providers" will survive and prosper in the newly competitive world of global telecommunications.

We have also described integration at the process level and the systems level as essential to achieving "lean provider" status. We gave you a glimpse of the standards world (which can really be confusing) and pointed out all the places where the industry could resist movement toward an environment that supports service management excellence. We made it clear (we hope) that service providers can't look to their telecom or computing suppliers to pave the way for them—on the contrary, suppliers will need to be led!

If you've just about decided that the job is too big, and you're about to dust off your resume, it's time for some good news. Kenichi Ohmae, formerly chair of McKinsey & Company's Japanese offices, wrote that "Companies are just beginning to learn what nations have always known: In a complex, uncertain world filled with dangerous opponents, it is best not to go it alone."[1]

The good news is that the service providers of the world are not left to "go it alone." Together, and with their equipment and computing suppliers, they are tackling some of the hardest integration problems head-on, and they are making progress. In this part of the book, we review progress that the industry at large has made to solve problems. In particular, we tell you what one organization, the Network Management Forum (NMF), has been doing for the past several years, the problems it is now pursuing, and how you can take advantage of its work.

Although other organizations are also of interest to service providers, the NMF has turned its focus toward removing barriers to service management excellence, addressing the need for industry agreements in areas no other group seems to be tackling. It also happens to be the place where we have both invested a great deal of personal energy since 1987.

The NMF was originally conceived in 1987 by Keith Willetts as a way of energizing the industry worldwide to back common and consistent approaches to interoperable network management. Teaming with John Miller of AT&T, Willetts worked with an extraordinarily dedicated team of people, including Elizabeth Adams, to develop the forum concept that has now become a model for cross-industry collaboration.

In July 1988, in a blaze of flashbulbs at the Plaza Hotel in New York, and at simultaneous launches in London and Toronto, the Forum was born with eight founder companies. One day, we will write another book about the enormous fun and headaches that setting up and managing a major international consortium can induce. Suffice it to say that the character and energy of the NMF was molded in those early days and has kept the Forum growing, delivering, and relevant.

The people involved with the outstanding and continuing success of the NMF are too numerous to mention, but we couldn't let the opportunity pass to mention a few individuals who made the concept work and whose dedication kept it energized: Ian Sugarbroad from Nortel, Leighs Church from Amdahl, Robert Montgomery from Telecom Canada (now Stentor), Dave Mahler from Hewlett-Packard (now with Remedy), David Milham from BT, Makoto Yoshida from NTT, Robbie Cohen from AT&T, Bruce Murrill, who left BT to

---

[1]Ohmae, Kenichi. 1989. The Global Logic of Strategic Alliances. *Harvard Business Review*, March/April 1989, p. 143.

become the NMF's technical director, Jim Warner of DCA, who later became the NMF's marketing director, and Marianne Trudel Jenkinson of M.F. Smith. To them and the hundreds of other professionals from the 30 countries in which the NMF operates, we thank you for your help and friendship over the last few years.

Perhaps we're biased, but we think the NMF is a great story. We hope you will, too.

Chapter

13

Introducing the NMF

The Network Management Forum (NMF) appears to be the only worldwide industry group in which public service operators, enterprise managers, and their suppliers have joined forces to solve shared problems to do a better job of service management. Although it had its beginnings in solving network management problems, it is now moving fairly aggressively to address service management issues, building upon its base and extending the value of earlier agreements while evaluating ways of exchanging information that are less tied to telecommunications standards and more tuned to advances in computing technology.

It is a very pragmatic organization that is closely linked to business issues. It does not develop standards—in a way there are already too many—but rather acts as a forum for reaching business agreements and selecting and integrating appropriate standards to implement those agreements. It provides clear, no-nonsense guidance for people whose job it is to buy technology within a service provider company and unambiguous implementation details for those who have to supply it.

Today, the group consists of more than 160 companies in nearly 30 countries. All of the world's major service providers, computing companies, and telecommunications equipment suppliers contribute to the work of the NMF. It remains true to its original roots in that it continues to focus on interoperability, but its scope and focus have evolved over the intervening years as service providers have undergone their own metamorphoses.

The Role of the Service Provider within the NMF

Within the NMF, the initial interest of the service providers was from groups responsible for developing management capabilities for the enterprise market. In that role they were competing head-to-head with computing companies and equipment manufacturers to produce "manager of manager" systems to help customers handle their multivendor network world. As they began to develop managed services and feel the icy winds of competition, the service providers' role within the NMF broadened.

As service providers started to focus on what it meant to achieve service management excellence, their role within the NMF, vis-à-vis the telecom and computing equipment suppliers, changed from that of competitor to that of customer. Service providers began to realize what they needed to do to compete in their industry, which was to add value to equipment bought from the telecom suppliers that could be managed in a common way and to buy computing platforms capable of supporting their management applications in a fully integrated systems environment. In essence, the service providers—which are the largest users of managed equipment, management systems, management applications, computing platforms, and related tools—began to assert their buying power.

From the standpoint of the NMF, this transformation has meant that what was once clearly a vendor-led consortium (with service providers, computing vendors, and equipment suppliers vying equally for a share of the enterprise management system market) became an organization driven by the needs of the largest users of computing and telecom equipment—the service providers.

The Current Focus in the NMF

Today, program direction is set by the service providers, and suppliers have clearly taken on the role of helping the industry to change. The NMF's current focus on meeting the needs of service providers and other large enterprises has earned its recognition as the place where the most difficult management problems are addressed and the only international consortium where both service providers and their suppliers can work together to solve shared problems. The NMF has always benefited from having a balanced view from three critical industries (network operators, telecom equipment suppliers, and computing vendors) involved in roughly equal measure.

Groups led only by vendors often come together to fight a shared foe, and initially they may make fast progress. But eventually, competitive issues get in the way of reaching any agreements other than bad compromises, and the value is diminished. Groups consisting only of users are usually long on demands but short on solutions because they don't have the benefit of participation from the very companies they hope to influence. When a group is equally represented by buyers and sellers, somehow problems keep getting solved. How many suppliers are willing to sit around the table with their largest customers and their largest competitors and be the only one to block forward progress?

Although the combination of buyers and sellers is a must, it's not enough just to have one representative of each company around the table. Many different constituencies exist within service provider, computing, and telecom equipment companies. Unfortunately, these constituencies speak different languages. Some speak in business terms, others speak technology; some speak communications while others speak computing. In reality, all of these languages need to be heard in order to achieve end-to-end process automation across the boundaries between processes, between service providers and their customers, and between service management systems and networking technology.

It is with these constituencies in mind that the Forum's programs are structured the way they are and why the outputs of the NMF look the way they do.

Programs for Integration: SMART, OMNI*Point*, and SPIRIT

There are three key program activities that address the needs of the multiple constituencies within service providers and their suppliers, called SMART, OMNI*Point*, and SPIRIT. These programs address the differing needs of process-level integration and systems-level integration.

SMART—The NMF's Service Management Automation and Reengineering Team

SMART focuses on reaching industry agreements at the business or process level, with particular emphasis on information that must cross corporate boundaries. The outputs of SMART are sets of agreements that define the information exchanges between customers and service providers, between multiple providers, and between service management systems and the underlying networking technology used to deliver services.

These agreements bring value even when an automated intercompany link is not required. An example is an agreement on a common set of terms and definitions to be used in constructing service level agreements. For the most part, however, the agreements are struck specifically because there is a need to automate a linkage. An example is the need to exchange trouble tickets or billing records automatically and in a common way among companies. Those SMART agreements that are intended to result in an automated link are augmented by technical detail as part of the NMF's OMNI*Point* program.

OMNI*Point*—Open Management Interoperability Point

In the OMNI*Point* program, the emphasis is on selecting and integrating standards and technology to be used in automating key processes involved in the development, delivery, and management of communications services. The outputs of OMNI*Point* are complete implementation agreements called OMNI*Point* Solution Sets and Component Sets.

OMNI*Point* Solution Sets are implementation agreements designed to solve a single, specific management problem. Solution Sets include a description of the business problem to be solved, a complete implementation agreement that details the information to be exchanged between two systems, and a reference list of the underlying standards needed to support Solution Set development. Examples include automated trouble-ticket exchange or the ability to dynamically change bandwidth provided to a customer.

OMNI*Point* Component Sets define the underlying technologies used to implement Solution Sets, as well as tools to make it easier to develop systems or applications. Examples include management platforms, applications programming interfaces, and protocol interworking specifications. OMNI*Point* Component Sets help providers and their suppliers implement reusable technology that will support multiple solution sets.

In general, OMNI*Point* Solution Sets and Component Sets represent a concrete and precise way for buyers of management systems and manageable products to specify their requirements to suppliers. Equally, the OMNI*Point* documents can be used to negotiate automation agreements between service providers and their customers or between multiple service providers.

SPIRIT—Service Providers' Integrated Requirements for Information Technology

SPIRIT addresses the computing needs of service providers. The output of SPIRIT is a set of procurement specifications, agreed upon by major service providers from Asia, Europe, and North America, that spells out both the specifications for a large-scale mid-tier computing platform and the specifications to be met to integrate across the mainframe, mid-tier, and desktop environments. The SPIRIT platform is generic in that it can be used to support any application, including business applications (e.g., payroll and accounting) as well as service and network management systems specific to the communications industry. The SPIRIT procurement specification is available from the NMF and X/Open Company as a stand- alone document that includes specific procurement advice for service providers.

The three programs link together in a cascade as shown in Figure 13.1. They are discussed in greater detail in the next three chapters.

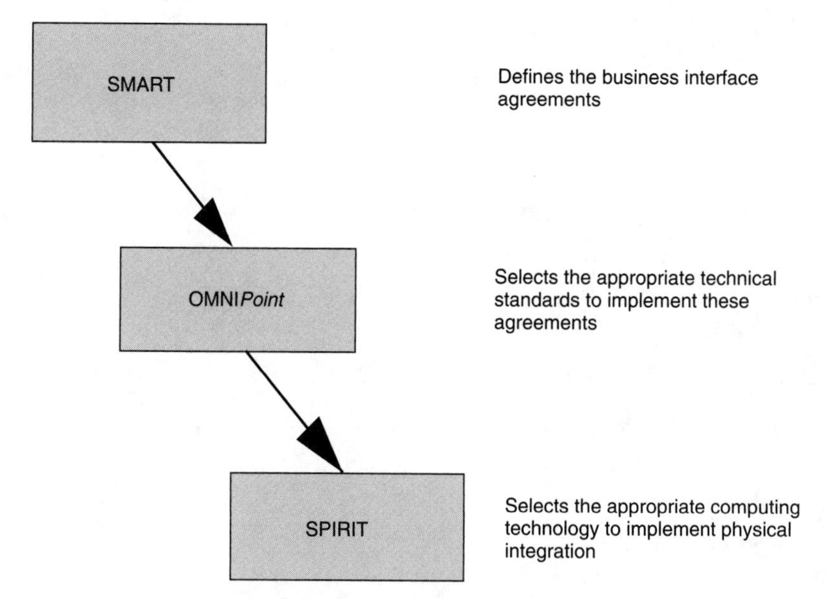

Figure 13.1 NMF's programs for management integration.

Chapter

14

Getting SMART about Business Process Agreements

Chronologically, SMART is the NMF's newest program, but in terms of output, it is most appropriate to put SMART at the front of the line. Here is where the service providers—including traditional PTTs, global carriers, new entrants, and enterprise managers—decide where they need to work together to improve flow-through of key service management processes.

Prelude to SMART—the RECAP Project

Leading up to the creation of the SMART initiative, a small NMF-sponsored team led by Elizabeth Adams conducted a requirements capture (RECAP) project to determine whether service providers saw the need to reach intercompany agreements and whether there was general agreement on areas of highest priority. Service providers from different parts of the world were interviewed, and each interview involved a cross section of people from each company. This work effort had several positive outcomes.

- It served as a way to begin to identify people within the service provider community who understood service management and the need to improve process flow-through.
- It confirmed that there was a real and growing need to focus on service management issues and that this need was not being satisfied elsewhere on an international basis.

- The information gathered as a result of the interviews served as the basis for developing a document later published by the NMF as "A Service Management Business Process Model."[1]

A common model: Overcoming the proprietary language barrier

Nearly all the people who were interviewed as part of the RECAP process agreed that they had critical needs that required industry agreement to solve, and they had similar ideas about priorities. But when they were brought together to reach agreement on the work that should be done, they found it impossible to communicate. Any discussion of common interfaces necessitated discussion of processes that each company defined differently and that represented the company "jewels." With no common process model, it was impossible to reach agreement on requirements.

The RECAP team took a major detour at that point and concentrated on producing a simple model that captured the key processes used by service providers to develop, deliver, and manage their service offerings. As input, the team obtained proprietary process flowcharts from multiple companies under nondisclosure agreements. The challenge was to look for common threads across all companies, which turned out to be a fairly simple task. Much more difficult was settling on a few key process names and descriptions that everyone could live with.

The resulting business process model does not match any one company's process names, nor are any two service provider companies organized to deliver the functions in the same way. Some companies combine many functions into a single customer-contact function, while others maintain more traditional organizational separations. Regardless, the point is that each service provider is able to make the connection between the terms used in the common model and the terms used within their company. And, more important, with a common model and set of terms, service providers can now converse with their competitors when necessary to do business without revealing how their companies have chosen to structure themselves internally.

A Look at the Service Management Business Process Model

Figure 14.1 shows the simple model that has now been accepted by service providers as a fair representation of how they do business. There are a total of 16 processes identified in three categories: customer care, service development and operations, and network and systems management. Of these 16, only the ones marked with a small square are viewed as potentially involving an intercompany process link, either because of a customer-supplier relationship (one provider orders service from another to complete an

[1]Network Management Forum. 1995. A Service Management Business Process Model.

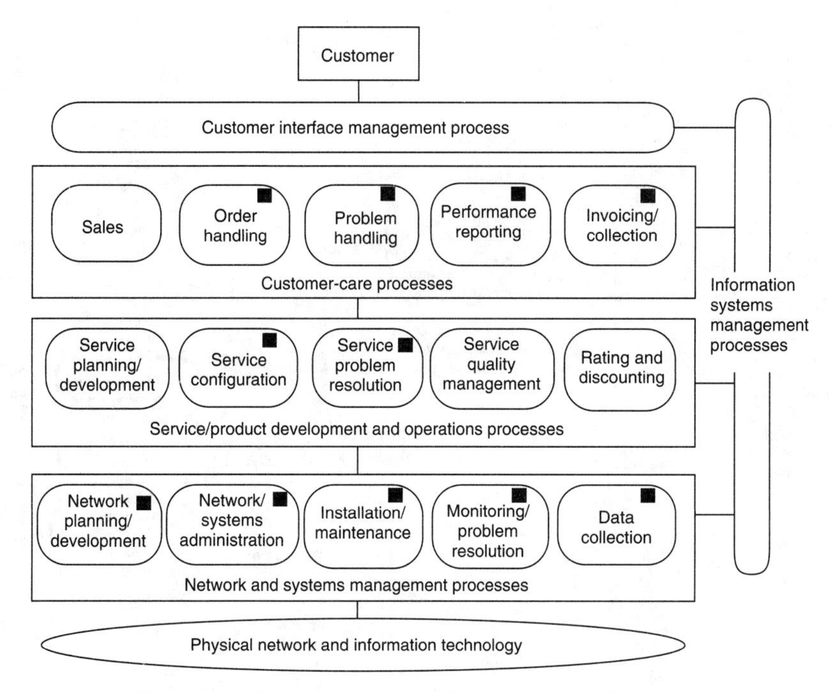

Figure 14.1 The Service Management Business Process Model. *(Network Management Forum. 1995. A Service Management Business Process Model)*

end-to-end service to the customer), because of a contractual arrangement (e.g., a third party performs billing functions for the service provider), or because of a dependency upon third-party equipment suppliers.

In addition to this overall model, more detailed descriptions of each of the service management processes (both customer care and service development and operations processes) were developed, showing not only the functions performed within each process but the inputs and outputs of each. In this way, it is possible to see the relationships between processes and to understand how an agreement made in one place might impact data in other processes. Figure 14.2 shows the expansion of the service configuration process as an illustration.

The individual process flows highlight possible areas where an intercompany interface might be required, by showing in bold any places where the flow of information (either on the input or output side) might involve organizations external to the service provider's company. In the service configuration process (Figure 14.2), the input of capacity data will traditionally come from management systems that are provided by a third-party supplier. So, although the equipment might well be on the service provider's premises and maintained by the service provider, any changes to the data format generated by that equipment (or its management system) will in-

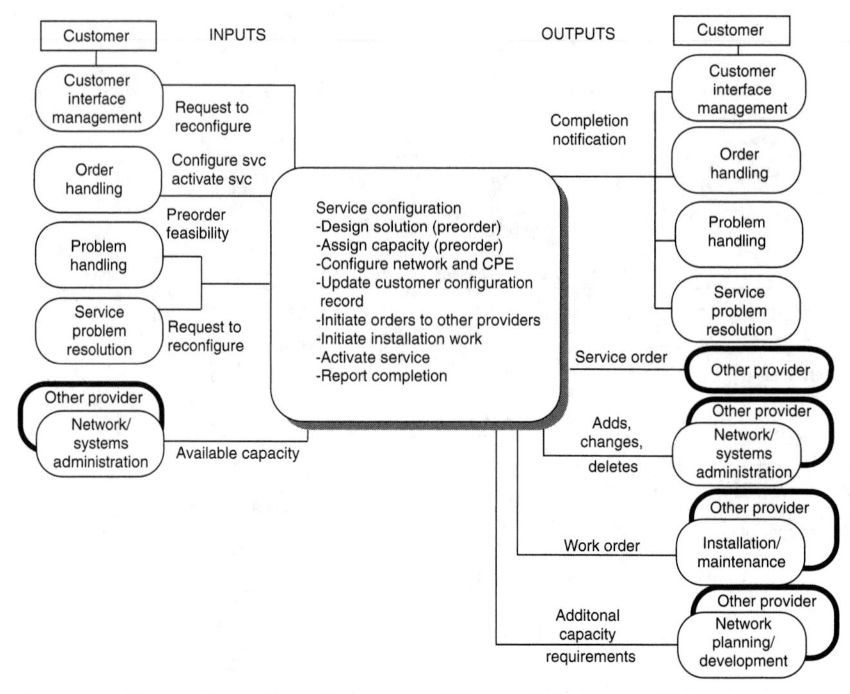

Figure 14.2 The service configuration process. *(Network Management Forum)*

volve reaching agreement with the supplier. As another example, outputs of the service configuration process (such as a work order) might be transmitted to another company that agrees to perform installation of local-access plant on behalf of the service provider. To the extent that the company performing this loop installation function wishes to receive work orders in a common format, or that service providers are likely to work with multiple such outside installation companies, an industry-wide agreement would probably be preferable to forging individual one-on-one agreements.

As service providers reach agreement in some areas, they might well find that additional interfaces could benefit from industry-wide agreement beyond those now marked as potential targets of emphasis. Certainly, companies that form very tight alliances will find that the need for intercompany linkages will extend beyond the generally accepted areas of intercompany interaction highlighted in the business process model.

Where the Barns Are Being Burned Down

As we see in the next chapter, industry agreements have already been reached in a number of areas as part of the NMF's OMNI*Point* program. These have tended to focus on the service level to network or element level interfaces (vertical process integration) involving the monitoring of alarms or the configuration of equipment. Under the SMART initiative, the focus of agreements is now being broadened to address the horizontal process integration issues that reside within the service management level.

There are a few key areas in which service providers have agreed that work needs to be done as a high priority. Norm Meyers from GTE, who helped to shape the SMART activity, would say in Texas-speak that these represent places "where people's barns are being burned down." Translation: These are areas for which there is a sense of urgency to reach agreement.

Four areas have been identified thus far, having to do with order handling, problem handling, performance reporting, and billing. The intent of each of these initiatives has not been to automate the process itself but to focus on automating key interfaces with a customer, another service provider, or a supplier. Within each of these processes a specific interface has been identified as high-priority. These work initiatives are reviewed individually.

Order handling

There are a great many opportunities to automate the order handling process, because this one process connects (either on the input or the output side) with all of the other customer care processes and several of the service development and operations processes, as shown in Figure 14.3.

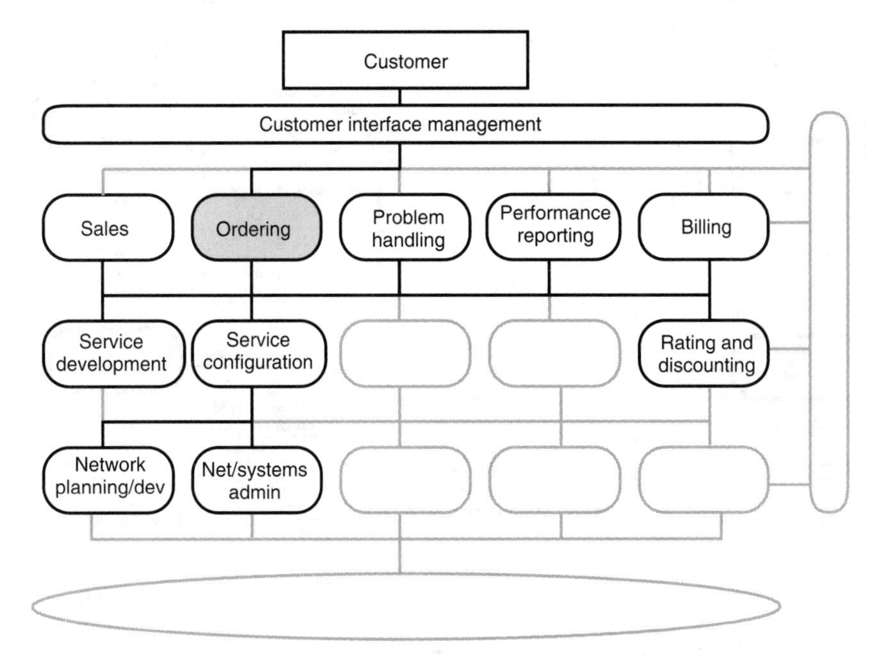

Figure 14.3 Showing the links between the ordering process and other processes. *(Network Management Forum)*

Order handling involves interactions with both end customers and service providers that are next in the service delivery chain. A current industry priority is to automate the link between service providers, which will help them do business with alliance partners and others as they enter the global marketplace.

Initially, the focus has been on defining a common way to track the status of order progression, because a relatively small investment allows a significant reduction in risk when dependent upon other providers to complete an end-to-end service. Figure 14.4 highlights the specific inputs and outputs of this process that have the greatest priority for agreement.

Automating the agreements they reach in the area of order handling will let service providers:

- Communicate with companies in different time zones (especially helpful when the gap is eight or more hours).
- Quickly access up-to-date information without having to find a responsible person via telephone or fax.
- Take action quickly if they see that a critical date may be missed.
- Keep their own end customers informed of order progression status with greater confidence that the information is correct.
- Reduce the cost of manually ascertaining the information.

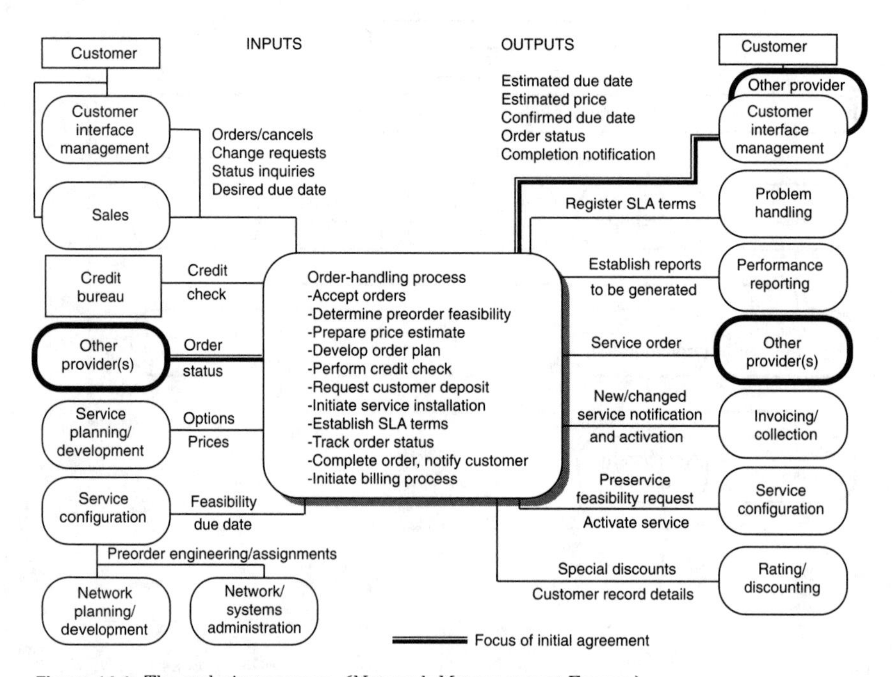

Figure 14.4 The ordering process. *(Network Management Forum)*

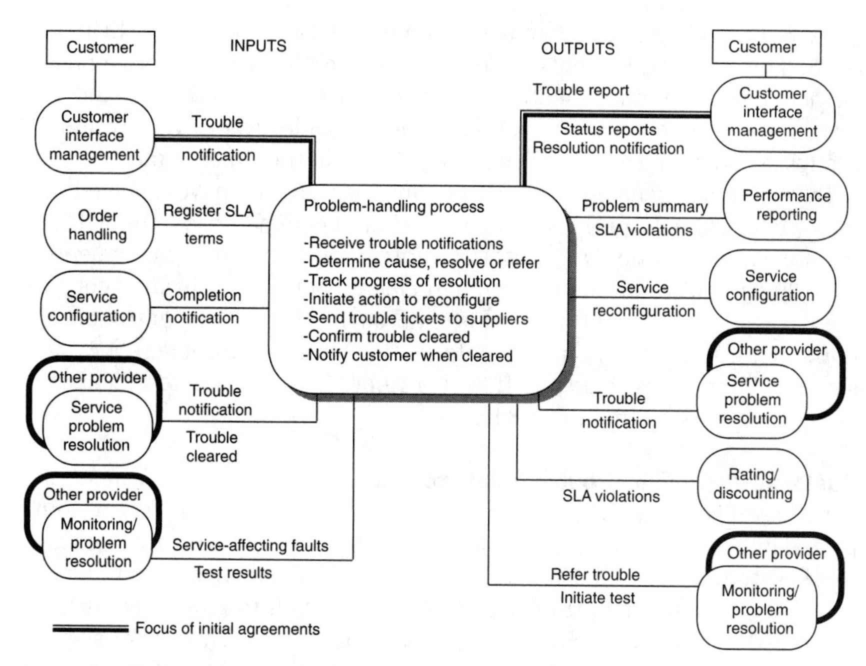

Figure 14.5 The problem handling process. *(Network Management Forum)*

Once this initial agreement is implemented, the work may be extended in several ways, such as:

- Exchanging not only order progression status but initial orders, as well, at least for the most common service types, such as leased circuits.
- Giving customers electronic access to order status information using their own internal service management systems.
- Extending automated links from the ordering process to other key service management processes, such as service configuration and billing.

Problem handling

Two interface agreement activities have been initiated in the area of problem handling. Both are based upon the same process shown in Figure 14.5 and both are focused on the link from the service provider to the customer. However, in the first activity, the customer is another service provider, while in the second activity, the customer is an enterprise customer.

Provider-to-provider trouble ticket exchange. This interface agreement activity builds on work done earlier by the NMF and Committee T1 as part of OMNI*Point* 1 and takes advantage of implementation agreements forged by the U.S.-based Electronic Commerce Implementation Committee (ECIC) of ATIS. The ECIC agreements developed between local and interexchange carriers have been implemented as part of a larger "electronic bonding" effort aimed at automating paper agreements made over several years in groups such as the Ordering and Billing Forum in the United States.

But whereas specific electronic bonding agreements have involved tailored, bilateral data profiles implemented on a common base, the SMART initiative aims to identify a subset of information that can be implemented globally without requiring tailored agreements between individual companies. The initial focus of this effort is to automate the tracking of trouble reports, but it may grow to include information to open a trouble report on a small number of basic services that represent high-volume interactions between providers around the world.

Customer-to-provider trouble ticket exchange. There are a significant number of requirements for common interfaces that might ultimately be supported by an automated problem handling interface between providers and their customers. The first priority is to provide customers access to status information on pending trouble reports, after which the interface may be expanded to address the ability to open a report as well. This work has benefited from work already done at the provider-to-provider level and has involved selecting appropriate subsets of trouble ticket information that are important to be exchanged with enterprise customers. Additionally, this effort has focused on defining the implementation constraints that must be met to achieve wide uptake by enterprise users and the applications developers that sell help desk software. It is expected that customer premises-based systems will need a solution that is easier and less costly to implement than the TMN-based solution used for provider-to- provider exchange. At least two alternatives, the use of Internet simple network management protocol (SNMP) and the use of the remote procedure call function of distributed computing, are possible candidates for endorsement by the NMF in an implementation agreement.

Performance reporting

The performance reporting process is shown in Figure 14.6.

The performance reporting initiative within the NMF addresses three business problems that are a little bit different in scope from the other SMART activities. Here, automation is not the key driver, at least not initially. Rather, the emphasis is on improving the human process flow-through that frames the customer-provider relationship for measuring and describing service performance results.

Nonstandard SLAs. Service providers are increasingly establishing service level agreements (SLAs) with their customers, which spell out the contrac-

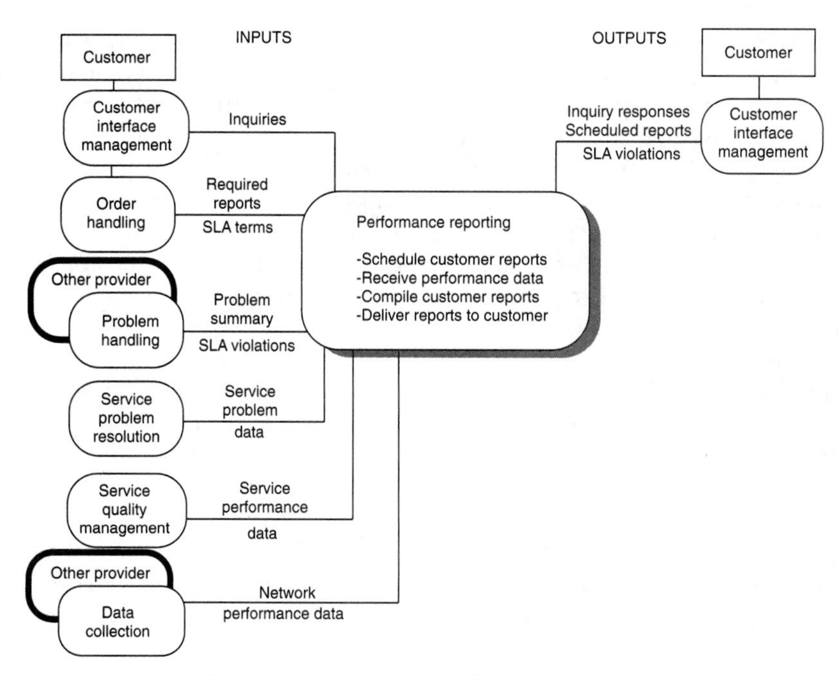

Figure 14.6 The performance reporting process. *(Network Management Forum)*

tual obligations of the service provider to the customer and usually involve some sort of rebate if the agreement is not met. Items covered under an SLA can and will vary depending upon the market objectives of the service offering. For example, a service provider might offer its residential voice-service customers two service options: standard service, under which service outages are fixed as soon as possible, but with no specific guarantees, or a premium service, under which, for an additional fee, the customer is guaranteed to be back in service within a defined period—otherwise, rebates apply.

SLAs can be an important service differentiator, but they can also be difficult to administer. In fact, the specific instance that prompted the formation of this SMART activity is that very large data-network customers were demanding (and getting) individually tailored SLAs. As there was no industry-standard list of available measures from which to choose, these SLAs contained not only different combinations of measures but such a wide range of measures that the service provider found it very difficult to keep track of the different agreements operationally or to make the connection between events in the network and particular SLA provisions.

Nonstandard definitions of terms. The second problem is that there are no common definitions for any of the measurement terms commonly in use in SLAs today. A favorite measure is availability, yet there are more measures of availability in use in the industry than there are service providers, since equipment suppliers have their own view as well. Coming to terms (no pun intended) with the terminology used in reporting performance is very important in managing customer expectations. If the service provider defines availability as the amount of time a particular circuit is working (however impaired it might be), and the customer thinks it is a measure of how many times the end-to-end service accurately and quickly transmitted data from one place to another, the service provider will probably lose this customer the first time a rebate is demanded and the customer is told "we met the terms of the SLA."

Service providers that are heavily dependent on other providers for network connectivity have a strong interest in defining common terms. Without them, it is impossible to construct a view of the overall performance of the service provided to the end customer. The use of standard terminology needs to extend from the service level right through the network and into the elements, and a connection will eventually need to be made between all of these levels in order to effectively administer SLAs. For now, the chief concern is to reach agreement among service providers on terms to be used when reporting service performance to end customers.

Relevance of information to the customer. The final goal of this effort is to reach an understanding with customers about exactly what information is really important to them. The telecom industry is noted for providing information that is available rather than what is needed by the customer to manage his or her business. How many customers are interested in knowing how many errored seconds occurred within a four-hour interval? Do they really care what the mean time between failures might have been for all users of a certain service, or are they only interested in knowing whether *their* service was the one that failed?

Getting this kind of requirement input from customers is extremely difficult. For one thing, there are different constituencies within the customer's organization—some that want technical information, others that only want to know when their applications cease to work, and some in between. Also, customers have other, more pressing problems in managing their networks, and, as we pointed out earlier, most enterprise customers see their service provider link as the least of their worries.

Billing

Another SMART initiative involves billing. This is an area in which service providers have no shortage of problems, most of them involving internal difficulties of achieving end-to-end process flow-through when their billing systems are ancient and inflexible. But even in the area of intercompany agreements, the members of the NMF acknowledge that they want to work on all three critical points of interface: with equipment suppliers' networking equipment, with other providers, and with their customers.

With so many challenges facing them, the service providers have agreed to make their first area of focus the feed of billing information from the suppliers' element management systems to their network and service management systems, as shown in Figure 14.7. In particular, they will focus on the billing of broadband services—something all service providers are struggling with as part of introducing new multimedia services. If this initiative can make rapid progress, several companies have mentioned the desire to apply agreements to cellular services.

Working SMART

Each SMART activity goes through several steps. They are designed to make certain that business needs are met and that solutions are practical for the people who will be asked to implement them.

Step 1: Define the business-level requirements

The most important part of the SMART process is to state the problem clearly, in business (not technical) terms, identifying specific benefits that are expected to be achieved by reaching agreement on a common interface. This also includes describing the exact information that is to be exchanged between systems, the actions each system is expected to take, and expected responses to actions. These requirements will generally include both a static and a dynamic information model to show how the information will pass back and forth. In addition, these requirements describe the environment in which the interface is expected to be implemented.

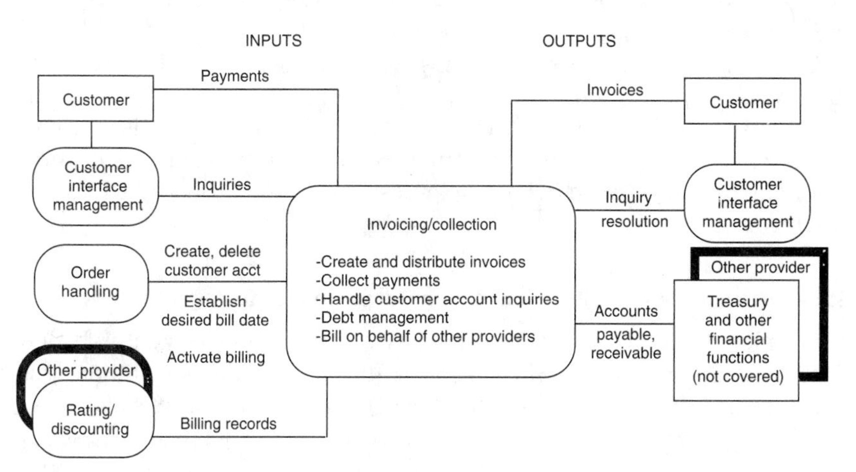

Figure 14.7 The billing process. *(Network Management Forum)*

For example, the customer-to-service provider trouble ticket exchange is likely to involve PC-based systems on the customer's side and mid-tier or mainframe systems on the service provider's side. Considerations need to be given to the cost and ease of implementation for both parties, as well as the levels of performance and security that will be required.

Step 2: Develop one or more draft specifications to meet the requirements

It is at this point that people knowledgeable about standards and systems development become involved. In other words, this is where draft OMNI-*Point* specifications are developed. Depending on the environmental requirements, the team may choose to take advantage of communications agreements already developed, or they may feel it is necessary to look at other solutions. For example, as mentioned earlier, the customer-to-provider trouble ticket initiative will explore three technical solutions—one based on the CMIP protocol, one using SNMP, and one using remote procedure calls (RPC). As a result of its work, the team might decide that more than one solution needs to be endorsed: one for customers whose needs more closely match those of the service providers and one for customers looking for an easy-to-implement solution that will adequately support low volumes of exchanges.

Step 3: Implement one or more of the draft technical solutions

To make certain that SMART activities are not initiated except when there is a well-defined and urgent business need, a ground rule was adopted that says SMART agreements will not be considered complete until and unless they are implemented by the participants. In practical terms, new work is not initiated unless at least two companies are ready to put their money on the table and implement the agreement.

At the time of writing this book, this step is being tested. The customer-to-provider trouble ticket interface is being implemented in a trial involving the U.S. Defense Information Systems Agency (DISA), Bell Atlantic, and GTE. Telegenics, developer of communications interoperability software, and Remedy Corporation, whose help-desk software is used by DISA, are also involved.

Step 4: Develop an OMNI*Point* Solution Set to finalize the agreement

The implementation experience gained as a result of Step 3 is only of value if used. Depending upon what is found during the implementation test phase, changes to the agreement may be needed or new solutions may need to be tried. At some point, a formal technical specification is developed in accordance with the OMNI*Point* program (reviewed in the next chapter).

How SMART benefits the industry

Participation in SMART presupposes that a company has recognized the need to establish common interfaces with its trading partners. No other international organization seems to have taken on this mission, and no other organization is likely to have both the membership reach and the technical knowledge to address these issues effectively. The companies that participate in SMART activities cite several significant benefits from their involvement.

Resource sharing: Do the job for less money. Individual companies faced with negotiating interfaces with trading partners face an uphill battle and high risk of failure. But by contributing a relatively small amount of staff resource to SMART activities, companies can achieve the equivalent of agreements that would otherwise have required many times the resource. The result (if past experiences hold true) will be more complete and useful because more viewpoints will have been considered.

International reach: Solutions that can be applied anywhere. SMART, as part of the NMF, is an international activity, and agreements arising from SMART are contributed to all appropriate international standards bodies as part of the OMNI*Point* program. While many regional efforts are now tackling one or more specific interprovider electronic-communication issues, worldwide partnerships are forcing a wider view. SMART is able to take advantage of any work already done regionally and advance it within the international community. This is particularly appropriate for the growing number of global alliances that can benefit from an international "neutral" focus rather than a regional or proprietary view.

Learning firsthand what service providers need

Suppliers of telecommunications equipment, computing equipment, or software that choose to participate in SMART learn what is needed by the world's service providers and have the opportunity to educate the providers regarding what is feasible to deliver.

Chapter

15

OMNI*Point*: From Business Agreements to Technical Solutions

OMNI*Point* has evolved over the years and has changed shape to reflect the changing nature of business problems faced by the industry. Its chief focus is to forge links between buyers and sellers of management systems and technology and thereby reduce industry's costs and lead times. More specifically, the focus is to arm service providers with down-to-earth business tools that will enable them to provide clear guidance to their suppliers of management functionality. Suppliers, in turn, get a common story from a range of customers, with clear specifications attached, allowing them to feel more able to develop generic product solutions rather than custom products. Without this customer-supplier link, widespread, cost-effective, and timely management solutions are not possible.

How OMNI*Point* Solutions are Developed

The outputs of the NMF's OMNI*Point* program are the actual implementation specifications that make it possible for developers to implement the SMART business agreements in a consistent way. In this phase of the NMF's work, several activities take place.

Selection of standards

Standards and specifications are selected that best meet the business and systems-environment requirements of a particular interface (with the objective to reuse the same underlying management standards wherever possible). These include everything from management protocols to management functions to object definitions.

Where there are holes in the standards, help is requested from the appropriate standards group. If nothing can be done in the time needed, the NMF may do work to fill gaps, although these days that is rare since generally there are more standards available by far than have been used. The one area in which new specifications are likely to be developed is in object definition, although even there the NMF's work generally extends base objects that already exist, such as the generic equipment object.

Profiling of selected standards

Standards invariably permit options in their implementation so that they can be used to solve multiple problems. However, when trying to solve a specific problem, decisions must be made with regard to each option—is it required or not allowed? If interoperability is to be achieved, the implementation details need to be absolutely precise so as to leave no room for interpretation on the part of the developer. This process is known as producing a "profile" of the standard—a specific implementation agreement that is capable of being tested.

Ensuring architectural consistency

To keep development time and costs low, every effort is made to reuse basic building blocks. When it is necessary to add to the underlying "toolset" to meet a new requirement, care is taken to integrate the added specifications with what has already been agreed upon by the industry. In this way, it is possible to chart a reasonable development path from what exists today toward new management technologies. Several specific activities help ensure consistency:

- Development and maintenance of an "architecture" for management systems to show the relationship among the various management standards and specifications.
- Development of interworking specifications required when two management "environments" need to exchange information and each environment uses a different management system approach.
- Development of migration guidance, required as implementation specifications are updated over time to reflect needed changes. As a general rule, updates are not made any more frequently than two years apart in order to give developers and procurers a reasonable return on earlier investment.

Packaging the outputs to aid procurement

The most tangible value of OMNI*Point* is found through its outputs—procurement and development packages called Solution Sets and Component Sets that deliver the technical agreements. These "sets" are a new and welcome addition to OMNI*Point*, since its outputs used to be much more difficult to use successfully. The aim in producing these sets is to help users of management systems (service providers and others) get exactly what they want from suppliers or negotiate a tight intercompany automation agreement with another provider or customer.

OMNI*Point* Solution Sets

A Solution Set is driven by a specific business need. Usually this need involves the exchange of management information between service providers, between a service provider and a customer, or with a supplier's equipment management systems. These types of agreements constitute the vitally important core of TMN, called Q3 and X interfaces. This exchange, or business agreement, can be made at the service management level, the network management level, or the element management level and can address either horizontal process integration needs (such as between two service management systems) or a vertical integration need linking the management layers. Solution Sets are single-purpose solutions, although they make use of a common infrastructure, and the agreements made in one case may be extended to apply to another problem.

Figure 15.1 shows examples of the kinds of information that might be exchanged at the service, network, and element levels. At the service level, horizontal interactions involve such things as service orders, trouble tickets, performance reports, and invoices. The vertical interface from service level to network level is used in the provisioning and monitoring of a customer service, whereas the interface from the network level to the element level involves the provisioning and monitoring of network connectivity on behalf of all service users.

The agreements being developed as part of the SMART program capture several examples of the types of agreements that constitute service-level-to-service-level Solution Sets, which are horizontal process-integration agreements. These are trouble ticket exchanges, performance reporting agreements, and order status exchange agreements, as shown on the top tier of Figure 15.1.

Vertical agreements, linking the service, network, and element levels, are equally important to achieving service management excellence. And while many vertical integration agreements involve proprietary, internal process linkage, service providers have found cases in which industry agreements are required—usually because the network or element level management systems

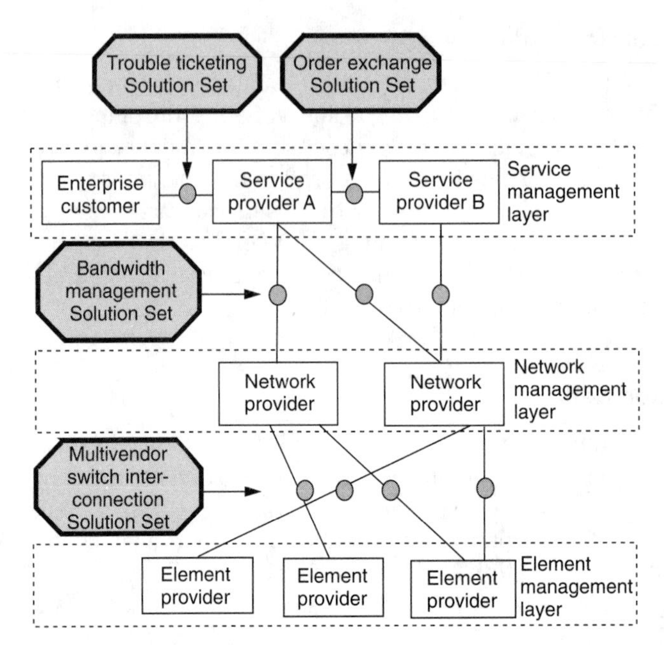

Figure 15.1 Solution sets are implementation agreements that solve specific problems. *(Network Management Forum. 1995. Implementing OMNIPoint.)*

are provided by multiple vendors. Solution Sets of this type have been developed to address the management of switch interconnection, the administration of customer service features, the monitoring of a LAN, generic alarm management, and the management of a variable bandwidth service.

Multinetwork bandwidth management: A Solution Set example

There are a multitude of technologies available to be used by service providers in the construction of communications services, and customers rarely care which technology is employed as long as the service they receive meets their requirements. Variable bandwidth services, for example, might be provisioned using ATM, SDH, or TDM technology, or a combination of the three. In addition to multiple technologies, it is likely that the various networks used to provision a variable bandwidth service may be supplied by different vendors.

In an environment in which each type of technology from each vendor responds to different command sets, automation of this reconfiguration function is impossible. Manual reconfiguration is the only alternative, and that means the process of making a change is too slow to be of value to the customer. This case shows a clear business need for a common way to communicate specific information.

The service provider would like to be in a position to make changes at the service level and have those changes be automatically implemented at the network level. Even better would be to give the customer the ability to reconfigure its bandwidth on demand, say to accommodate a sudden video conference, and to have that service command result in the necessary changes at the network level. The specifications of this Solution Set make that possible in a consistent way across multiple network types provided by multiple vendors, as shown in Figure 15.2.

Although the focus of this Solution Set is the service-level-to-network-level interface, sufficient information is also provided to make clear what must be implemented in the element management systems to support the higher-level interaction.

What is not so clear from looking at this particular Solution Set is the number of other activities the service management system must take care of. For example, before permitting a customer to increase bandwidth, the service management system needs to verify that the customer is authorized and that capacity is available. It needs to initiate a change to the billing record so that the customer pays for the higher bandwidth use, and so on. Over time, as service providers face these difficulties and as the service provider industry is further fragmented, some of these automation challenges may have industry solutions.

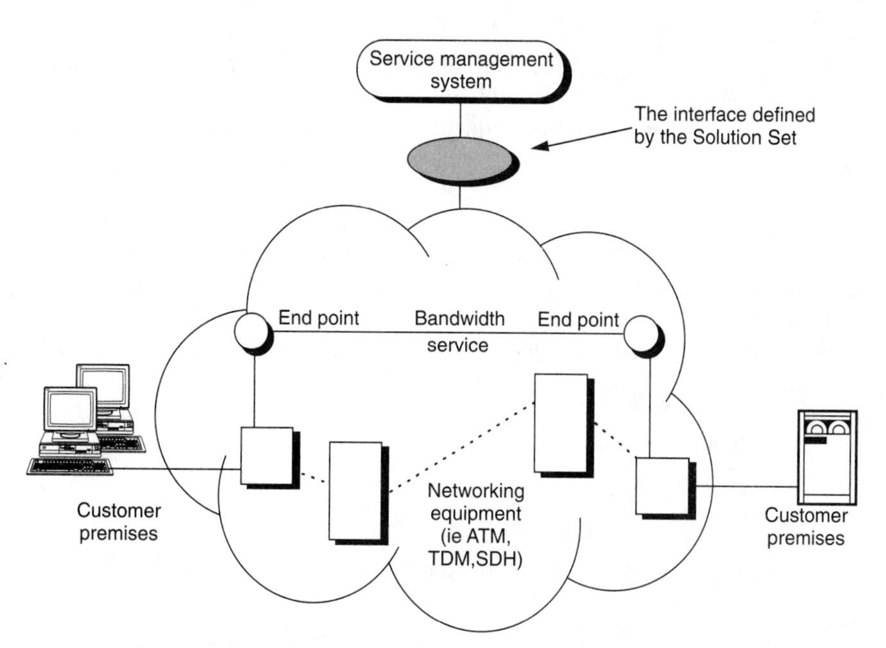

Figure 15.2 The Multinetwork Bandwidth Management solution set. *(Network Management Forum. 1995. Multinetwork Bandwidth Management Solution Set (NMF SS201).)*

The contents of a Solution Set

Every Solution Set is a complete implementation agreement. Each set includes a product descriptor and one or more technical specifications. It may also include tutorial requirements statements or "white papers" to aid in using the Solution Set.

The product descriptor is a brief summary of the Solution Set. It describes the business problem being solved, the scope of agreement, the benefits to the target audience, and its applicability in different business contexts. It describes the environment for which the Solution Set is designed and addresses any migration issues in the event that similar functionality might have been offered in an earlier OMNI*Point* release. Finally, it states the requirements for conforming to the Solution Set and offers procurement advice, including any dependencies or additional requirements to be satisfied to implement the Solution Set. The product descriptor is intended to give sufficient information to determine whether or not it is desirable to require compliance from suppliers or other providers.

The technical solution to meet the needs of the product descriptor consists of detailed specifications that, if implemented, provide the specific management capability. At the heart of the technical solution is a document known as an ensemble.[1] An ensemble includes a management information model that describes the information to be made available from a set of resources (e.g., equipment, circuit) and the functions that can be performed on those resources (e.g., configure them, collect alarms or performance data from them, etc.). It also includes one or more scenarios to illustrate how a function can be performed using the information model and to show the sequence of activities, boundary conditions, and steady-state conditions that can be viewed across the interoperable interface.

So that the Solution Set can be tested to ensure that the supplier has met the criteria laid down, ensembles contain reference implementation conformance statements (ICS) that specify conformance requirements in a tabular format. The supplier that wants to demonstrate conformance to a Solution Set should complete these ICS tables, indicating which options and capabilities have been implemented. Conformance requirements for a Solution Set are organized as follows:

- Managed object classes.
- Function profile requirements.
- Management communications profile requirements.
- Implementation conformance statements proformas.

---

[1]Credit for the ensemble concept goes to Jock Embry of Opening Technologies.

Thus a Solution Set has two primary, interlinked parts: a product descriptor that gives enough information to support a business decision to proceed with implementation and a technical solution containing all of the details necessary to achieve management system interoperability. Each Solution Set includes conformance criteria so that the implementation can be checked, making contractual agreements (such as payment) straightforward. The aim is to simplify the process of buying and selling the various technologies that make up a service provider's armory of management systems so that they can be integrated. Integrated management allows process flow-through, which in turn delivers better service management, resulting in better, faster, and cheaper services!

OMNI*Point* Component Sets

An OMNI*Point* Component Set defines a reusable implementation of management technology that enables Solution Sets to be implemented. Component Sets are of most interest to developers of management systems that wish to buy underlying platforms and tools that will interoperate with other suppliers' products.

There are likely to be a small number of Component Sets relative to the number of Solution Sets over the life of the OMNI*Point* program. The business goal, in fact, is to build many different applications-related solutions upon a very contained base of management-interoperability facilities to minimize the total investment and maximize the value of investment in such technology.

Component Sets reference three types of specifications: management platforms, interworking capabilities, and development tools such as applications programming interfaces (APIs). Among the most interesting to service providers may be the TMN Basic Management Platform Component Set, explained in the following paragraphs.

TMN Basic Management Platform: A Component Set example

Service providers have been asking their computing suppliers for "TMN platforms" for some time, and some suppliers have started to deliver products with that name. However, because there has not been a specific definition of what must be included in such a platform, these products vary widely in their usefulness. This variation has made service providers nervous about investing in an "unknown" as a base for mission-critical management applications.

The TMN framework and body of related recommendations represents an enormous and very complex challenge that will take time to evolve into mature technologies. However, there are elements of TMN—a sensible "starting point" subset of the many TMN standards—that deliver immediate value. The OMNI*Point* TMN Basic Management Platform, illustrated in Figure 15.3, captures this starting point and provides the level of detail needed to implement certain key aspects of the TMN architecture.

Figure 15.3 shows that implementation of the TMN Basic Management Platform Component Set is dependent on the implementation of the CMIP Communications Component Set. For data communications, developers are advised to check the NMF's SPIRIT specifications, which provide several options that support interconnection of platforms in a common way. TMN platforms that comply with this Component Set are capable of providing the basic support needed to implement any Solution Set built upon the TMN framework and associated standards.

By requiring compliance to the TMN Basic Management Platform, there should be no doubt about what is to be delivered. Suppliers may choose to provide additional functionality, but they will not have the option to ignore the basics. Over time, the NMF members may choose to define additional TMN functions that can be applied to the Basic TMN Management Platform in a common way to extend its usefulness.

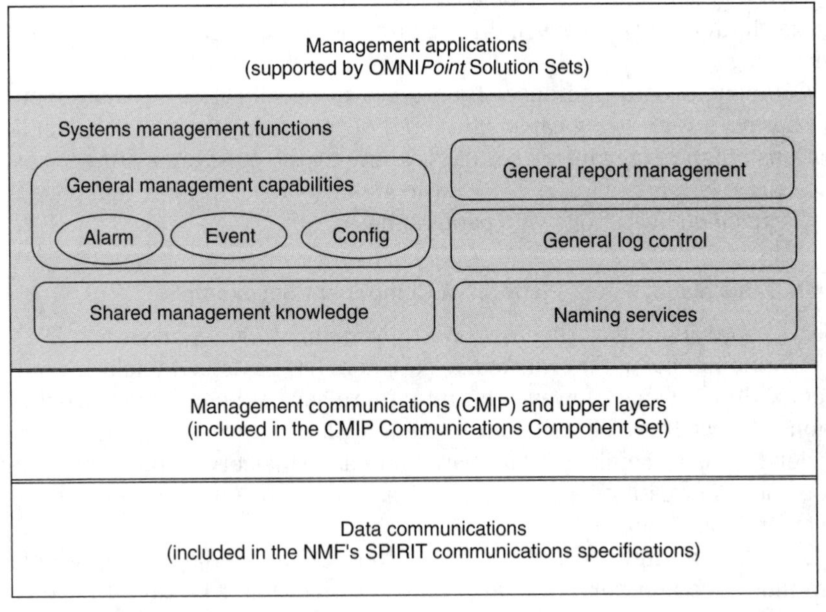

Figure 15.3 The shaded area shows the scope of the TMN Basic Management component set. *(Network Management Forum. 1995. TMN Basic Management Platform Component Set (NMF CS302).)*

Interworking Component Sets

Anyone who follows the details of management technology is familiar with two communications protocols—CMIP and SNMP—that support command and control functions (things like alarm monitoring and configuration) with varying levels of capability. In addition to these two fundamental management building blocks, advances in object-oriented technology are introducing higher-level techniques for communications.

Linking these two protocols is important to service providers, since they will probably use CMIP to manage "backbone" networks and sophisticated services but may need to extend their management reach onto the customer's premises, where SNMP is widely used. This is particularly true of service providers that offer customers an end-to-end service to replace their private networks or that manage their customers' private networks through an outsourcing arrangement.

While individual standards exist for each of these protocols, there is no interest on the part of the (often rival) standards-setting organizations to make them work together, although developers find that problems of interworking are better tackled collaboratively than individually. The NMF serves as a linking agent here, and it has developed a CMIP/SNMP Interworking Component Set. In addition, the NMF is addressing the linking of object definitions when stated in terms specified in the OSI model, the Internet model, and the CORBA (common object request broker architecture) model from the Object Management Group.

Contents of a Component Set

Similar to Solution Sets, Component Sets include a product descriptor that summarizes the functionality represented by the Component Set, the benefits of using it, examples of how it might be used, conformance requirements, and any procurement advice that is appropriate. In addition, the Component Set either includes specifications needed to implement the Component Set or references standards and specifications produced by other organizations. White papers or technical reports, including specifications for migrating from earlier-referenced OMNI*Point* standards where appropriate, are also included, as are conformance requirements and procurement advice.

Relationships among Solution Sets and Component Sets

Among the things the NMF has tried to do in organizing its OMNI*Point* sets is to minimize repetition of information in every set. For example, several of the Solution Sets require the use of the CMIP communications protocol and a number of basic management services. Although these references could have been repeated in every Solution Set, the decision was made to separate out anything used repeatedly, making it a Component Set.

Within the Component Sets, there is also a layering of information based on the procurement patterns that are likely to develop. For example, the TMN Basic Management Platform Component Set contains all of the functional specifications (the management services and common services) needed to support most management applications, but it does not include the communications protocols or APIs that might need to be applied to the platform. Instead, it references other Component Sets for these purposes. These separations were made because the NMF believed there was a market for each of the components, and they wanted to make it possible for as many vendors as possible to claim compliance with OMNI*Point*, while at the same time leaving no doubt on the part of the buyer exactly what compliance to a specific set means.

Version Control of OMNI*Point* Sets

Each OMNI*Point* Solution Set and Component Set is maintained separately, so that it can be updated when there is a good business reason for doing so. Sometimes a set includes the use of a standard that is still in draft or preliminary form. As the standard advances through the standards process, changes may be made that enhance the capability of the standard. At some point, companies that have implemented the set (containing the draft standard) might find that it makes sense to migrate to the later version of the standard. At that time, the set will be updated and migration guidance will be provided to aid developers in upgrading existing systems to reflect the change.

Just because a standard becomes "final" does not necessarily mean that service providers or suppliers will find a business need to change their systems to incorporate the change. When there is a strong case for doing so—as evidenced by the willingness of the NMF member companies to update a Solution Set or Component Set accordingly—then the change will be made and that specific set will be up-issued. Through version control, everyone who plans to implement a particular management capability will know exactly which version of what standards are needed in order to interoperate with other companies that have also implemented, and that is really the issue when systems need to communicate with each other.

Industry Problems Addressed by OMNI*Point*

Three key problems exist in achieving interoperability that OMNI*Point* aims to correct.

Effectiveness of procurement

A large number of standards exist in the area of network and service management. Buyers have no good way to relate individual standards to specific management problems and so are likely to include such vague phrases as "compliance to TMN" in their requests for proposals (RFPs). Because of the lack of specificity, suppliers are forced either to work with the buyer to gain more specific requirements or to claim compliance at a general level. In either case, the buyer rarely gets value from procurements, and identical procurements from multiple suppliers yield widely differing results. Certainly, interoperability among management systems is not achieved.

Cost of procurement

Service providers that attempt to solve the first problem (lack of specifics) undertake elaborate engineering efforts to produce detailed requirements. This adds direct cost and delay to the procurement process. In addition, product costs are higher than necessary because suppliers are forced to customize solutions to meet each service provider's unique requirements and are unable to spread the cost of development across a wider market. The lack of a common direction from buyers, at a level of specificity that is meaningful, means that suppliers are hesitant to invest in major developments.

Reaching peer-to-peer or subcontractual agreements

Service providers are increasingly developing strategic relationships with other providers to extend their global reach. In addition, they realize that to satisfy the requirements of multinational customers, they must be able to do business with many providers that might not be strategic partners. Reaching agreements on how to exchange orders or how to manage in-service performance is difficult and time-consuming and sometimes pits partners against each other during negotiation. Most providers are finding that it makes better business sense to employ industry agreements for exchanging data and to concentrate limited staff on improving internal systems and methods in order to differentiate their offerings.

How does OMNI*Point* address these problems?

- OMNI*Point* focuses on areas in which there is agreement on the need for common interfaces at the boundaries between companies; in other words, where companies have determined that it is in their best business interest to automate links with customers, other providers, or supplier-provided equipment.
- OMNI*Point* creates a clear relationship between standards and specific business problems to be solved by providing Solution Sets and Component Sets. By buying or developing against these sets, it is possible for companies to get to the heart of what they need without wading through tomes of standards that might not be required to solve their problems.

- OMNI*Point* provides the specificity required to turn base standards into very precise implementation agreements so that developers are not required to interpret the standards or to reach separate agreements with other developers on a one-on-one basis to achieve interoperability. This specificity can be the basis of contractual agreements constituting clear payment milestones.
- OMNI*Point* provides a good indication of current priorities, since OMNI*Point* Solution Sets and Component Sets are developed only when a clear business need is identified and "champions" are willing to do the work necessary to take a business requirement through to a full implementation agreement. That's important, because it means that developers can implement standards in order of their importance and not "wholesale." What's more, they can be reasonably certain of a market for their product following development.

How OMNI*Point* Needs to Evolve

Trying to determine how industry players will react to each other over time is always a bit risky, but everything in the NMF's history suggests that service providers and their suppliers will continue to drive for the development of more and more standardized implementation agreements. Today, the link between an ordering system and a billing system may be proprietary, and many companies would probably say that they need to remain that way in order to keep some competitive advantage. However, as soon as these functions begin to cross company lines—as links need to be forged between communications providers and information-content providers, for example, or as providers segment themselves into value-added providers, network operators, and infrastructure providers—agreements will need to be reached.

As agreements continue to reach more closely into how processes link together at the service level, the NMF's OMNI*Point* program will need to take advantage of different technologies to serve as an implementation base. To this point, with most agreements bubbling up from the networking concerns of suppliers, the use of ITU "TMN" standards such as the CMIP protocol have met the need for intersystem communication. But now that higher-level service management issues are beginning to be addressed, the industry will find that the basic infrastructure now used by management systems will need to be augmented through the use of generally available distributed computing technology and an advanced object-oriented environment.

The early stages of this realization are already in evidence as the NMF looks for the right technical solution to be used to interconnect customer systems with service providers' systems. The obvious choices from the traditional management technologies—CMIP and SNMP—might not be as effective in this application as other common computing techniques. Although the total number of different infrastructural technologies must be kept to a minimum, it is important to move forward and to take advantage of "disruptive technology" when it can make service management process integration possible.

It is encouraging to see the computing suppliers helping to pave the way toward future environments and looking for ways to link the more traditional management systems with the environments of the future. Taking pains to protect investment that has already been made while still moving forward to take advantage of new ways of doing things is one of the priorities of the NMF's members.

Chapter

16

Underneath It All: The Systems View from SPIRIT

The need to achieve service management excellence (balancing the three-legged stool) has driven service providers to streamline their processes and create common interprocess links. In the same way, the need to achieve cost reductions, faster development cycles, and better overall control is driving the service providers' information technology (IT) professionals to change how they do business.

Within the NMF, information technology experts from major telecommunications and computing companies formed a team in 1993 called SPIRIT (Service Providers' Integrated Requirements for Information Technology). This team has developed a single set of requirements for a scaleable distributed computing platform that, among other things, is capable of systems-level integration.

The Business Drivers: A Process of Continual Discovery

When the team first formed, its primary motivation was cost, as shown in Figure 16.1. Service providers were, as a group and individually, paying too much for their computing software. First, they were all investing large numbers of personnel to develop detailed requirements used to guide purchasing decisions. Second, they paid custom prices for software because their requirements were unlike any other company's. Third, they paid enormous sums to keep their custom software maintained. Something clearly had to be done, as staffs were being cut and the custom software could no longer be effectively maintained or managed.

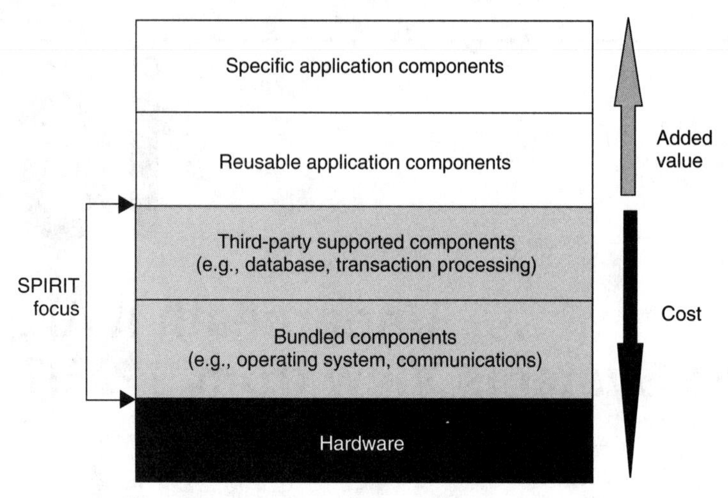

Figure 16.1 Whole-life cost reduction is a key business driver of SPIRIT.

The team's secondary motivation was the need for a distributed computing platform that actually met service providers' needs. None of the companies had been able to procure client/server software upon which they could run transaction-intensive, mission-critical applications. They despaired of ever influencing the computing vendors sufficiently to listen to and respond to their large-scale needs at a time when the demand for client/server for the small- to mid-range customer market was booming.

Third, the increasing pressure from business process reengineering efforts was making it clear that linkages between systems—systems integration—was essential to achieving process integration, as shown in Figure 16.2. It was not possible for systems even sharing the same base platform to interoperate, much less to achieve interoperation between the platforms of different vendors. Without solutions to this problem, the IT infrastructure would prevent service providers from making changes at the rate they needed to in order to stay competitive.

In the roughly two years it took for these service provider IT managers to accomplish their objective—to define a single set of specifications that would meet their needs—their priorities changed. By the end of their project, the need for integration had gone to the top of the list, no doubt the result of stepped-up activity in process-level integration since 1993. Costs, while still a factor, were no longer centered purely around purchase price and had broadened to reflect whole-life costs. In fact, a senior executive from one of the key companies involved confided that his company expected to pay more initially for what they had defined but that whole-life cost reductions and improvements in flexibility would more than compensate if the company's needs could be met.

The Key Players

The service providers that were behind the formation of SPIRIT included NTT and BT, with two other service provider "consortia"—Bellcore and the European Telecommunications Informatics Services (ETIS)—rounding out the group of four. NTT, in particular, took a lead role, since it had had success with its internal multivendor integration architecture (MIA) program in Japan and was anxious to see its benefits extend to open computing platforms.

All of the individuals behind SPIRIT had been active for some time in X/Open, participating as members of its User Council. Yet they were disappointed with the lack of agreement coming from the suppliers and felt they could apply more pressure from outside the computing industry. They considered forming another new consortium but rejected the idea, not only because of cost concerns but because of the lead time associated with starting such an organization.

The link to the NMF was a natural. From the standpoint of those already active in the NMF, the SPIRIT initiative meant an opportunity to address the systems integration problems that were plaguing management systems development. It also meant that experts from the IT side of the company could become involved to help understand and address the systems-level integration issues involved in achieving service management excellence.

Attacking the problem: A few ground rules

SPIRIT has one key difference from other NMF activities. The founders were very concerned that under normal NMF rules of approval, the decisions of the service providers might be reversed or delayed indefinitely if full board approval were required to complete the specifications, since as many as one-third of the NMF board members were computing suppliers.

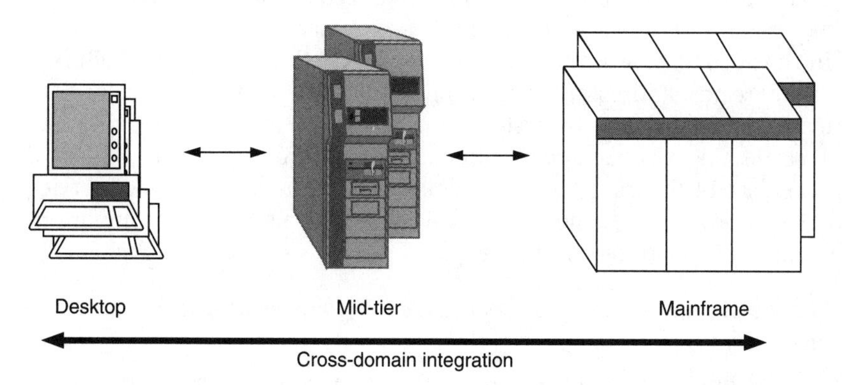

Figure 16.2 Integration across the three computing domains is another SPIRIT business driver.

To avoid competitive blockage from threatening the program, the NMF’s board delegated approval of the SPIRIT specifications to the SPIRIT steering committee—a group consisting only of service providers—although the board insisted that every effort be made to reach consensus with the suppliers that would form the bulk of the technical teams. As it turned out, reaching consensus was not a problem.

The SPIRIT steering committee, chaired by John Wright of BT with Kei Takagi of NTT as vice-chair, established a few ground rules of its own. Perhaps the most significant was the decision that specifications would only be considered for inclusion if they were available from at least two vendors (or were expected to be within six months). This meant not only that the team avoided getting caught up in endless debates trying to shape new technology but that the results of their efforts would be immediately useful. This pragmatism marked the SPIRIT work from the beginning, with obvious positive results.

A whirlwind effort

The initial work program of SPIRIT was completed in three phases, running from spring 1993 until the third quarter of 1995—only four months later than the original two-year target. Much of this was due to the extraordinary dedication of the SPIRIT technical director, John Van Meurs of KPN (the Royal Dutch PTT), who actually represented, and was funded by, ETIS.

The team made a fast start by putting existing procurement specifications on the table and identifying all the areas where agreement already existed. These included the NTT multivendor integration architecture, the BT open systems standards guide, the ETIS open systems components guide, and the Bellcore UNIX-based standard operating environment. The results of this sorting exercise marked the end of the first phase, resulting in a document called “SPIRIT Issue 1.0,” which itemized areas of agreement, and a structured set of priorities for the next two phases.

In the second phase, the team began to make a real impact. Technical working teams addressed everything from programming languages, to transaction processing, to systems management, to internationalization. The result was a set of specifications, published jointly by the NMF and X/Open as SPIRIT Issue 2.0, that spelled out the requirements of service providers for an open, general-purpose computing platform. Two important outcomes demonstrated its worth:

- Service providers began to procure against the specifications almost immediately.
- Several brave decisions involving transaction processing were made, breaking years-long deadlocks among the manufacturers and gaining acceptance by X/Open as part of its common applications environment.

SPIRIT Issue 2.0 did not achieve complete "plug and play," nor did Issue 3.0, which marked the completion of the specification, although more work was done in areas such as conformance, management, and security, and an effort was made to group the specifications into realistic procurement "bites" to aid in using the specifications. The group's objective did not extend to achieving full interoperability between platforms, both because there are still "holes" in the standards (and SPIRIT only selects from specifications but does not develop them) and because in some areas the industry is not yet mature enough that the service providers felt comfortable locking into a given approach. Still, the progress made in the area of systems integration represents an excellent start.

Making Use of the SPIRIT Specifications

We've covered many of the business reasons why SPIRIT is important, but it's not always obvious to people who hear about SPIRIT what new capabilities it represents or why it is important.

According to the SPIRIT platform blueprint:

> A software platform is a set of generic capabilities implemented in software which enable and facilitate the creation and operation of applications. A general-purpose software platform does not make specific assumptions about the nature of the applications that it supports. SPIRIT is not concerned with hardware architectures; indeed SPIRIT presumes that the software platforms it describes are implementable on a variety of hardware architectures.[1]

The software platform model

The SPIRIT software platform comprises programming languages and the following services:

- *Operating system services*, which manage the fundamental physical and processing resources of a given machine.
- *Management services*, which effect the changes of managed items on the platform.
- *Presentation services*, which act as the mediator between the system and human user-interface devices, such as display, keyboard, mouse, and so on.
- *Data-management services*, which manage persistent storage and presume some kind of data model; an example is a relational database-management system.

---

[1]Network Management Forum. December 1995. NMF Spirit Issue 3.0 SPIRIT Platform Blueprint.

- *Transaction services,* which coordinate resources to maintain transactional integrity over those data resources, applying updates in units called transactions.
- *Communications services,* which define how applications emit and accept protocols; for example, X/Open Remote Procedure Call Interface Definition Language.
- *Distributed services,* which facilitate cooperative processing between two platforms; for example, naming and distributed time service.
- *Security services*, which enforce security policies on data and processing objects on the platform and data objects exchanged between platforms.

The SPIRIT software platform model is shown in Figure 16.3.

SPIRIT not only selects from among many possible standards, but, where necessary, it also profiles those standards. There are two ways in which profiling is used. Specification profiles improve the likelihood of interoperability between systems by selecting which of the available options must be implemented. Component profiles, on the other hand, group sets of standards in ways that vendors might consider building products, so that it is clearer to the vendor which standards must be complied with, given the types of products it intends to produce.

The value of interoperability

People who come from the telephony world are used to having things that fit together—at least things that come from the same vendor! It would be odd, indeed, if two switches couldn't be connected to make a network. But in the computing environment, interconnection was an afterthought and is still not done very well. The telecommunications industry is often jeered at by the computing companies for taking so long to agree upon standards, but perhaps that's because until fairly recently computers weren't expected to

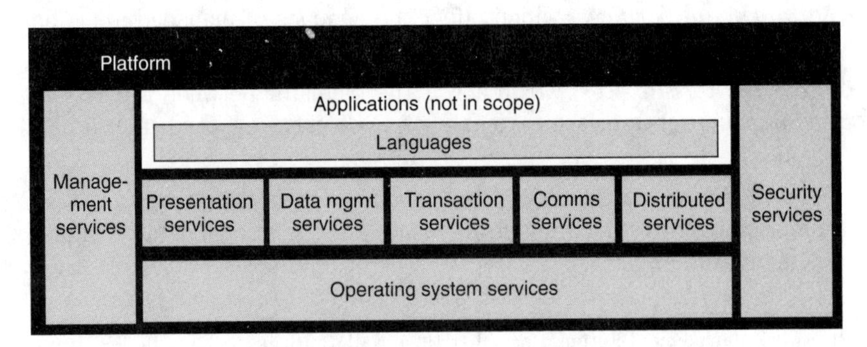

Figure 16.3 The SPIRIT platform model. *(Network Management Forum)*

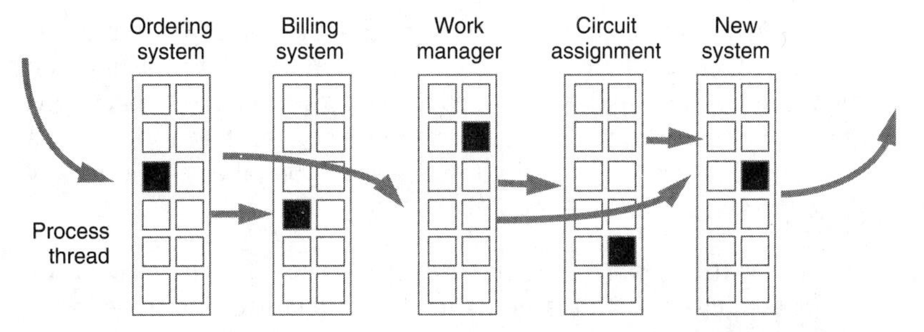

Figure 16.4 Process flow-through cannot occur unless the systems can interoperate.

talk to each other, so these companies couldn't appreciate the problem. Even today, software platforms claiming to be "open" cannot exchange management information even with like platforms, much less another vendor's "open" product.

People who come from the telephony world understand (and probably take for granted) the value of interoperability. When two systems can't talk to each other, they are worth much less than if they can. This ability to interoperate is among the best reasons for using SPIRIT, as shown quite effectively at an Interop Tokyo demonstration in 1994 and again at Telecom '95 in Geneva. Six different vendors implemented the transaction-processing standards adopted by SPIRIT, which were then invoked in an application developed by NTT to mirror a travel agency. Each system had a defined role (airline reservations, hotel bookings, routes and schedules, etc.) and contained data needed to complete an end-to-end travel reservation. Each was able to interact with the application, providing access to information as needed and making changes in response to commands.

Although this application was purposefully chosen to avoid a telecom bias (demonstrating the value of SPIRIT to any large user), it is similar to the kinds of process-integration applications needed to achieve service management excellence in the telecommunications world—linking the ordering system, the service configuration systems, the billing systems, and so on, as shown in Figure 16.4.

Without systems-level integration, such as demonstrated in the travel agency example of a SPIRIT implementation, these different service management systems will not be able to be linked, and process-level integration will not be possible at a cost that most service providers are willing to bear.

Dealing with legacy systems

Among the greatest concerns of the SPIRIT team is how to develop applications with the flexibility offered by distributed, mid-tier computers, while accessing data stored in centralized mainframe systems and using the desktop as the user-access environment.

One approach gaining increasing currency is the client/server/server architecture. This is an evolutionary approach that, instead of replacing the legacy mainframe in one maneuver, encapsulates the existing applications and data and positions the mainframe (or mid-tier system) as a server on a larger distributed network. Several approaches can be taken, but typically an evolution plan might have several phases:

- Phase 0: The status quo, i.e., existing centralized mainframes with legacy databases, applications, and large dumb terminal/cluster controller population.
- Phase 1: Dumb terminals are progressively replaced with PCs, and emulation or "screen scraper" techniques are used to improve the user interface. Mainframe applications are supplemented with local applications held on PCs or local servers.
- Phase 2: Centralized application development is frozen, and all new applications are on local servers but with access to centralized databases for data.
- Phase 3: The centralized applications are progressively dismantled by moving them to local servers.
- Phase 4: The central mainframe becomes a data server in a fully distributed environment.

There are a number of advantages to this encapsulation method. First, the risk of a major redevelopment of the mainframe application is significantly reduced. There have been many well-documented software disasters in which old legacy mainframe applications were replaced by new distributed ones only to find that operational performance or functionality were not fit for purpose. Encapsulation allows a step-by-step evolution, with each step being a small risk, since operations can always revert to the previous position if problems occur. Second, investment planning is improved, since equipment is not "cut over" in one massive effort but rather is phased in over a period of time.

Most important, users receive early relief in areas that trouble them most. The user interface is much improved at the beginning of the evolution cycle and can be tailored to match their own presentation layout, reports, and so on. Local applications development at the client end, rather than at the mainframe, short-circuits the mainframe bottleneck but does not interfere with the data integrity of the main database.

The advantages of a client/server/server approach don't come for free, and taking this path is not without its operational headaches. There are no perfect answers when trying to move from antiquated systems to something sleeker. But this option should certainly be among the tools considered by service providers as they attempt to deal with the legacy systems issue.

Making procurement easier: SPIRIT sets and conformance requirements

Similar to the way that OMNI*Point* has tried to make the buyer-seller link more clear, SPIRIT Issue 3.0 begins to provide guidance in terms of how certain specifications might be grouped for a specific purpose. Two types of "sets" are defined: system sets and component sets. The system sets are combinations of SPIRIT specifications for five typical platform configurations, including a manager platform, and two flavors each (transactional and nontransactional) of both a client platform and a server platform. For each set, certain specifications are mandatory, meaning that any supplier claiming conformance must meet those specifications. Other specifications are optional, which means that they are not required to be present, although if a supplier's product includes such an option, it must conform to the SPIRIT specifications.

These SPIRIT system sets may be further combined as follows:

- Nontransactional client + nontransactional server
- Transactional client + transactional server
- Manager + transactional server
- Manager + nontransactional client + nontransactional server
- Manager + transactional client + transactional server

SPIRIT component sets are consistent combinations of the SPIRIT communications- and management-protocol specifications. Five SPIRIT component sets have been defined: OSI upper and lower layers, Internet upper and lower layers, and DCE. The relationship between the SPIRIsT system sets and SPIRIT component sets can be seen in Figure 16.5.

In the area of conformance, SPIRIT Issue 3.0 makes more clear what is meant by conformance to its specifications. There are two ways that vendors can claim conformance to SPIRIT: "set-specific" or "general conformance." By using the sets, vendors can claim that they meet a defined subset of the requirements. By claiming general conformance, they are stating that to the extent that they offer functions or attributes for which SPIRIT specifications exist, they comply with SPIRIT. This is much harder to achieve, since SPIRIT covers a wide range of standards (more than 200 in all). It is unclear whether vendors will actually deliver complete platforms that comply in every respect with SPIRIT, but it is equally unclear whether service providers are likely to purchase whole platforms from vendors. The concept of sets holds promise for closing the buyer-seller gap by making it easier to use the SPIRIT specifications.

Meeting All of the Requirements

With the completion of the third phase of work, the SPIRIT program as originally proposed was finished, yet the SPIRIT program continues to function.

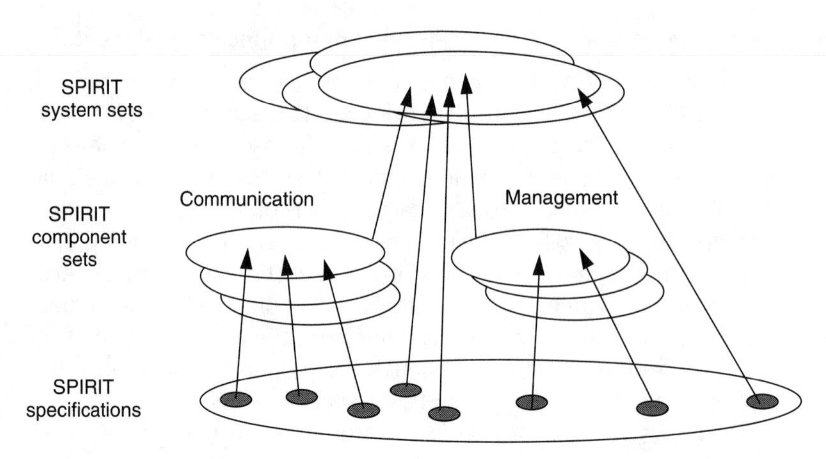

Figure 16.5 The relationship between SPIRIT system sets, component sets, and specifications.

The team has concluded that more work needs to be done, particularly in the area of systems integration. Having spent much of its effort through the third phase on achieving integration at the mid-tier, it is now focusing more closely on cross-domain issues, including more work on systems management. The team also wants to see that the specifications are understood, accepted, and used by service providers worldwide. It plans to find ways to keep the pressure on suppliers to comply through their products.

In addition, there are some service provider computing platform requirements, documented in a separate, NMF internal document, that have not been addressed by the SPIRIT specifications. As the technical work was proceeding with the help of the suppliers, a team consisting only of service providers worked to prepare a comprehensive set of requirements against which the final outcome of SPIRIT could be compared. This group, unlike the technical teams, was not constrained to include only those things that were already available in the industry, but rather was encouraged to give a broader indication to the computing industry of the needs that would have to be met with new technology.

Many of the requirements spelled out by this team were satisfied at the conclusion of the third phase. But some still remain as open issues and will guide future SPIRIT work. One area that continues to be most contentious is that of object-oriented technology. On the one hand, there is no doubt that such technology will play a part in future systems development—particularly management systems. On the other hand, many groups are already dealing with this new technology, and there is some question about what SPIRIT might contribute in this area.

Chapter 17

Taking Advantage of the NMF's Programs

The NMF's programs are tailored to meet the needs of different constituencies. SMART is aimed at capturing business agreements, OMNI*Point* provides detailed standards-based technical implementation agreements to satisfy the needs of procurers and developers, and SPIRIT addresses the problems of the IT environment. This chapter explains how the programs interrelate and the benefits of using the NMF's work.

SMART-OMNI*Point* Relationship

This relationship is fairly straightforward. As soon as the SMART business requirements are completed (and sometimes even as they are being developed), draft technical specifications are developed, which ties the SMART business people together with the technically oriented OMNI*Point* folks, as shown in Figure 17.1.

Although some of the OMNI*Point* Solution Sets had their beginnings before the SMART program was initiated, it is the NMF's intention that all new work involving process-level agreements should start out as SMART initiatives. In this way, a solid set of business requirements, representing the consensus of the parties involved, is developed before investing in technical solutions.

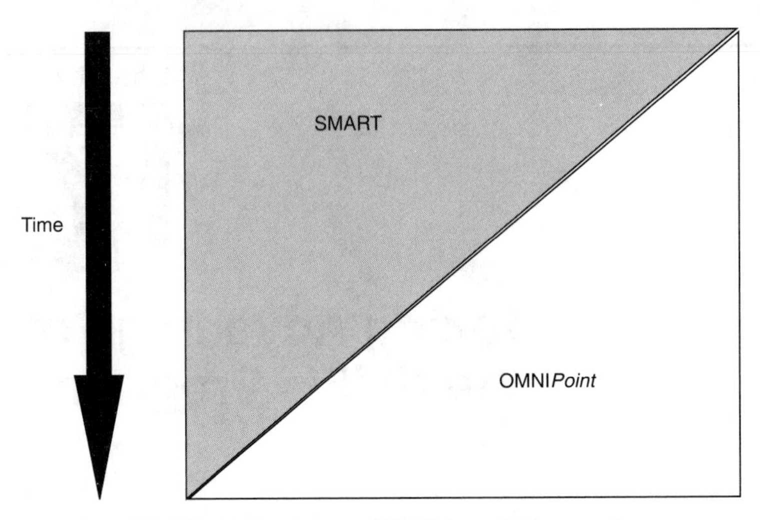

Figure 17.1 SMART initiatives become OMNI*Point* activities over time.

The OMNI*Point*-SPIRIT Relationship

This relationship is a bit more complex, particularly if a company needs to buy or build a management system platform. Look inside the OMNI*Point* Component Sets, and you find several that apply to the computing platform. That's just fine. But look inside the SPIRIT specifications, and you see some of the same things. Why does the industry need two packages that point to many common specifications?

Well, it depends on your perspective. If you are, first and foremost, responsible for buying or developing a management system, reference the OMNI*Point* specifications, since they are organized in terms that you will best understand. However, take note within the OMNI*Point* Solution Set and Component Set product descriptors of related SPIRIT references, and pass that information along to your IT supplier (whether that's someone within your company or outside it) to be certain that the underlying software platform is capable of supporting your requirements.

If your primary responsibility is buying or developing general-purpose management platforms (which can be used for all types of systems, including management systems), use the SPIRIT specifications as your main reference. But you might want to point out to your management systems clients that the platform you are going to purchase or develop implements certain of the OMNI*Point* specifications, which will be obvious from reading the SPIRIT specifications. Figure 17.2 shows these two different perspectives.

The reason for overlap is that process-level and systems-level integration are hard to separate. There is a point at which both need to address the same things. The management systems person needs to make certain not only that the application is developed in compliance with the information agreements (process level) but that the systems will be capable of communicating using certain management protocols and services. The IT person needs to make sure the platform he or she orders will satisfy a range of needs and will also need to make certain that the platform itself is manageable. So in addition to all of the general-purpose capabilities like transaction processing and programming languages that must be specified, the IT person will specify certain communications protocols and management services—the same ones specified as part of OMNI*Point*.

What's in It for You?

The service delivery chain encompasses relationships with customers, other providers, and suppliers. Each part of the delivery chain stands to gain significant business benefit from the work described in this part of the book, provided companies take the initiative to use the work.

Marc Malaise, a telecommunications consultant for Digital Equipment in Latin America, told us that his company's involvement in the NMF programs and his ability to take advantage of the NMF documents when talking with

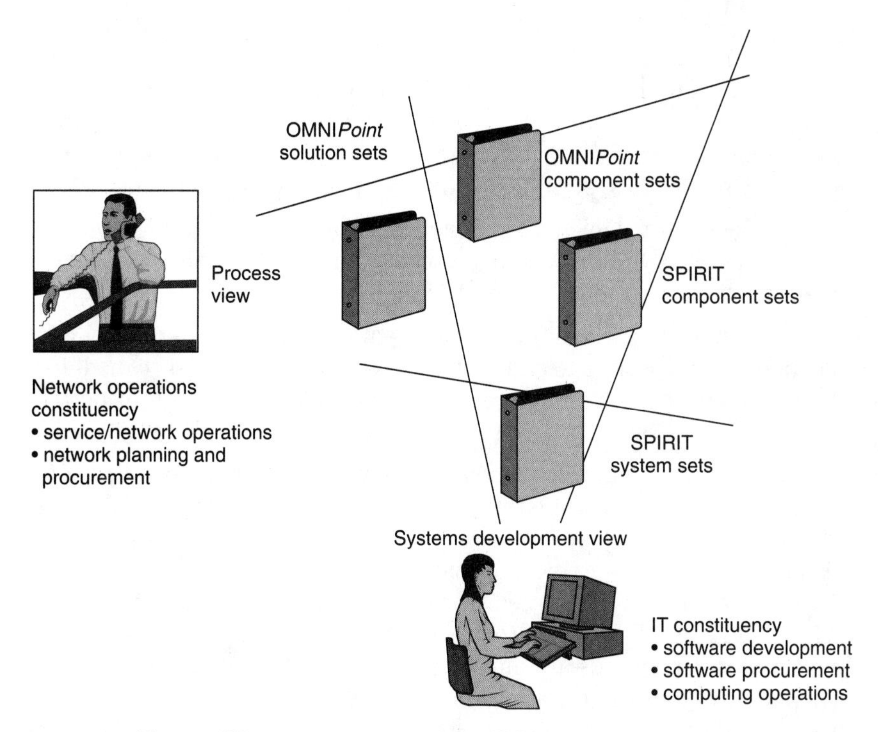

Figure 17.2 The two different views represented in NMF's outputs.

customers has added to his credibility, particularly when dealing with service providers that aren't yet as familiar with the issues of service management. Bernie Maier, manager of systems planning for Bell Atlantic, reports that many of his company's equipment suppliers continue to resist the move to standards. Being able to reference the NMF specifications gives him additional leverage in demonstrating that the standards are stable, that they represent what multiple service providers want, and that they can be implemented.

The good news: Benefits for all

Global enterprises, service providers, and suppliers of telecom and computing equipment all benefit from the agreements being developed. Global enterprises stand to gain:

- The ability to automate their links with service providers and therefore realize cost benefits and service improvements.
- The ability to employ the same procurement strategies as service providers for computing software platforms and managed equipment and therefore gain market leverage they might otherwise be unable to muster.

Service providers stand to gain:

- An environment in which critical interfaces can be common, to cut down on the costs of developing and maintaining one-on-one agreements with business partners or customers and to support process automation across company boundaries, even though internal processes remain proprietary.
- An environment in which it is possible to "buy in" a computing platform that is sufficiently robust and functional to support the types of management applications needed to stay competitive.
- An environment in which the management systems used to provide and guarantee good service to end customers are not required to be modified to accommodate different underlying technologies used to provide the service.
- A way to be more precise with suppliers when requiring compliance to standards such as the TMN-related recommendations from the ITU-T or the X/Open common applications environment, so that any two suppliers can meet the same stated conformance requirements, and interoperability (and value) can be realized. Without the precision offered by OMNI-*Point* and SPIRIT, service providers leave it to suppliers to determine which of the standards are important and how each of those standards should be implemented, with the result that costly procurements rarely result in useful solutions.

Telecommunications equipment suppliers stand to gain:

- Common requirements from their customers worldwide with regard to key data feeds, so that systems can meet requirements for critical standards even as suppliers differentiate their products through added functionality or better price or performance.
- An opportunity to significantly improve the speed with which new technology (such as ATM, SDH/SONET, or cellular systems) can be introduced by making certain it can be managed using agreed-upon interfaces.
- The ability to address management interface requirements as they build new systems rather than via more expensive retrofit procedures.
- A larger market for products, since their offerings will satisfy a wider number of companies' needs.

Computing suppliers stand to gain:

- The opportunity to meet a growing demand for software platforms that can meet the needs of their largest customers and that provide the basic capability to exchange management information in a fully interoperable way.
- Common requirements for key standards that their largest customers can agree upon, so that they can get on with the business of differentiating their products through ease-of-use, price and performance, or other factors.
- Insight into the uses of their computing products by the largest single-industry user of computing technology and with it the ability to respond to applications development opportunities.

The bad news: It's not a panacea

There are real limits to what the industry agreements described in this part of the book can do for you. In any of the three areas, there is a long list of management problem areas waiting to be addressed, and real solutions have been developed for only a small subset to date. That's one of the downsides of industry collaboration—it takes time to reach agreement, and the people involved all have regular jobs to do back at their offices, so actual time available to work on these issues is limited.

In addition, these agreements certainly don't do it all. Think of yourself as going on a journey. The NMF's programs represent the book of maps you'll need to get there, but you won't go anywhere if you don't put fuel in the tank, avoid crashing, and deal with unexpected detours that weren't marked on the maps.

In the struggle to improve your management processes and systems, you can look to industry groups like the NMF to provide a common architecture and a common language to be used to converse productively with suppliers. You can expect that agreements will be applicable worldwide. And you can expect (over time) that agreements will extend across the supply chain, provided the companies with a business stake stay actively involved to make certain progress continues to be made.

What industry groups can't and won't do is to shift your legacy base of systems or rewrite your management applications. They can't take labor out of your processes or guarantee that you will have the same operational result through implementing common agreements as your competitor; you have to correctly apply the specifications to achieve advantage.

Finally, industry groups can do a lot to inspire, expose, and embarrass suppliers into developing products, but at the end of the day, suppliers make the final decision based on many different reasons. The most important factor, of course, is buyer pressure, and that, too, falls to you to apply. The tools are there, but they need to be used in order to reap value.

Part

4

Getting to Excellence

No matter how much industry groups can help both service providers and suppliers in beginning to grapple with service management, the fact is, getting to excellence is an extremely tough feat to accomplish. Just getting started can be a major headache. Issues arise such as how to gain the support of senior executives to make enough investment in service management capabilities, how to gain the support of internal and external constituencies to make a change, and how to chart a reasonable course that will keep risks as low as possible and that will permit a phasing-in of changes.

Earlier in this book, we listed the things companies must concern themselves with when undergoing a major change, as in the case of today's tele-communications services companies. Out of the seven items listed, which included such things as articulating a vision and attracting the right skills, this book has been about only one of these: reengineering the operational processes and systems to improve service, reduce costs, and speed time to market—what we have called achieving service management excellence.

In the last part of this book, we focus on some of the key issues in implementing a strategy for service management excellence. Given the breadth and depth of the subject and the many disciplines involved, we inevitably can focus on only a few major topics that in our experience can be key to getting started and keeping up the momentum.

Chapter 18

Making the Business Case for Service Management Excellence

Senior executives of companies undergoing fast and major upheavals may be preoccupied with thoughts of mergers and acquisitions or assessing and responding to competitive threats and opportunities. They may find any discussion of process-level and systems-level integration somewhat boring if not irrelevant—especially if those "on the bridge" don't understand what happens down in the "engine room." Unfortunately, companies that are led by executives who are too concerned with weighty matters such as mergers and acquisitions to understand the engine room are probably doomed to be the prey rather than the predator.

We can all speculate what the future world of communications will look like. We outlined some of the key business drivers earlier, and some of the reshaping forces are becoming plain to see. In their excellent book *Competing for the Future,*[1] Gary Hamel and C.K. Prahalad introduce two key concepts that are important to understanding what organizations must do to win in the future. First is the simple concept of *foresight*—the ability of a company to see with some clarity the kind of industry that will emerge and how the company can position itself for a leadership position. Flowing from this idea is the concept of *competencies*—the capabilities and skills that an organization must grow or acquire to play a winning hand in that future.

[1]Hamel, Gary and C.K. Prahalad. 1994. *Competing for the Future.* Harvard Business School Press.

Both of these concepts are very relevant to service management. Our foresight shows a radically different industry than the one we have today:

- One in which complex value chains exist, and multiple players such as content providers, service providers, and infrastructure providers are involved in delivering services.
- A highly innovative industry with high product turnover and very short time to market.
- A price-competitive market in which service pricing is on a global basis without the distortions of cartels, monopolies, and interventionists.
- A market in which the customer is king and is treated as such with superb customer service.

Foresight is the easy part. What is difficult is to grow the competencies to deliver the future in the time available before that foresight becomes reality. It is timing that is crucial: Invest too late and your market might be gone; invest too early and hit all of the snags and teething troubles of any new skill or technology.

Service management excellence is the major factor in delivering the kind of "one-touch" industry vision we describe. It will be the major ingredient in success or failure, because the competencies required to deliver it take a long time to grow, especially if your starting base is a process infrastructure based on paper, faxes, and yellow sticky notes. The range of skills required and the degree of corporate cohesion takes years of hard work to achieve—they can't be bolted on when the company is in trouble and losing its market. As Hamel and Prahalad point out, markets are won or lost years in advance of when the effect shows up on the bottom line. Senior executives in an organization need to understand this cycle, but, unfortunately, too often they ignore the big underlying issues in favor of easy, short-term solutions. Executives in the U.S. car industry were still chanting the mantra "The U.S. consumer buys cars on style, not quality" years after the customer had started to vote otherwise.

So getting to excellence in the communications industry requires the attention not just of managers who have to implement programs but of the very senior executives who are paid handsomely to make big decisions. Without their active support and encouragement, any service management initiative (and the company) is unlikely to succeed.

Another current management writer, Peter Senge, in his inspirational book *The Fifth Discipline,*[2] sheds light on what comes next. Senge introduces a concept of "the learning organization"—a subject much harder to grasp and even harder to implement than either foresight or competencies.

---

[2]Senge, Peter M. 1990. *The Fifth Discipline.* New York: Doubleday.

Among other things, Senge points out that corporate leaders don't have all the answers and thus can't be expected to intuitively know that getting to service management excellence is any more or less important than myriad other issues. He advocates that organizations develop so that the entire management team listens and learns across the management structure as a whole—not just top-down, as in traditional organizations.

A lot of people at the operational level in communications companies can clearly see the future and what is required to beat the competition. What many cannot do is gain sufficient momentum and weight of management to make a major program of service management excellence really fly. And it's not just the established companies that suffer from such blind spots. A frustrated operations manager from a thrusting second carrier said to us "Look—this business is going to be won or lost on our ability to drive more services, faster and better, to our customers while continually driving down our costs so that we can beat the big guy on price. My problem is a board that is not investing for the future because of a very short-term approach to business."

We have written this section of the book to try to put some practical ideas down on paper about how to get a service management excellence program off the ground. It's like getting a flywheel turning—it requires an immense amount of effort to get started, but once underway, it gathers more and more momentum.

Getting started inevitably means putting together a business case, and that implies putting some hard numbers behind the theory that's been imparted in this book. It also means presenting the need for a service management strategy in terms that apply directly to the success of the business. If the business case for service management has not already been made in your company, it will need to be articulated clearly and succinctly before you have any chance of going forward. If it has already been made at very senior levels, sufficient to gain initial funding approval and direction of key resources, rest assured it will need to be made again, repeatedly.

Getting to service management excellence is a matter of continuous improvement in both large and small steps. Each improvement, no matter how large or small, requires investment of capital and personnel and involves risk. And as with any change, there will always be an army of people defending the status quo. So, developing and being able to cite a strong business rationale for proceeding is not a one-time thing; rather, it is a crucial way for the whole team to think about what each member is doing and why. Although senior executives are certainly one audience for the message, they are not the only one. Making changes to achieve service management excellence has the potential to affect nearly every employee in a service provider company and many of their suppliers. A great many people with individual fiefdoms to protect and years of expertise will need to be convinced along the way that the change is needed, and they, in turn, will have to be capable of convincing others down the line.

A strong and unshakable story that explains why change must be made is the most important weapon in the arsenal. Making that story simple enough that it can be repeated many times without distortion will make it an impressive weapon indeed. Making the business case for service management excellence is a nontrivial but absolutely crucial first step on the path to excellence. Those who have to deliver it must be classic champions who can lead, articulate, and convince the skeptical, and they must have hides like rhinoceroses.

Stick to the Basics

People who are very familiar with service management, who understand what it entails and the benefits of getting the three-legged stool in balance, often become frustrated that others can't see the whole picture. There is a temptation to dive into details, to unload real issues on people who cannot grasp the importance of the issue—not because they are obtuse, but because they haven't read this book! Seriously, understanding service management is not a casual undertaking, and it requires a view of the business that very few people have glimpsed.

In a recent meeting, an interesting discussion confirmed this point. Two people from service provider companies were trying to explain to two supplier types why it is so hard to do business with telcos. One of these service provider representatives—who has a technical background and is a recent convert to service management—acknowledged that in his company, counting himself, there were perhaps only 50 people out of 100,000 or more whose job assignments enabled them to see the end-to-end picture and appreciate how the pieces needed to fit together. His observation mirrors our own experiences, as well as the experiences of the NMF in initiating the SMART activities.

The only way to avoid frustration is to constantly remind yourself (or find a peer who can help remind you) that this is not an easy subject to understand and to keep it simple. Get the message down, and deliver it, again and again and again. It's easy to be fooled into thinking that people who live and breathe the TMN management standards have already made the business connection, but that's probably not the case. Equally, someone with profit-and-loss accountability for a product line may understand bottom-line results but may not see the link between achieving those results and achieving service management excellence as we have defined it. It's often as though this issue is so large that very few people can see it in its entirety—only the parts are readily visible.

But they can learn and become interested in service management very quickly if the subject is introduced properly—if you are able to hit the right buttons, which are different depending on the audience. Starting from the top—the profitability of the firm and its competitive position in the industry—is crucial to make a strong link between corporate success and the ability to achieve service management excellence. With senior executives, discussions of OAM&P systems (operations, administration, maintenance, and provisioning) will most probably cause their eyes to glaze over, and leading off a presentation to this group with an overhead entitled "The Path to TMN" is likely to lose you your best opportunity to make a sale.

Each constituency within the service provider or supplier community is likely to be interested in different aspects of the message. Some may resonate with the need to improve customer satisfaction results, others will be interested in hearing new ways to trim the budget or to shorten product development cycles. The same basic message is important to every group, although each will have its own priorities that should be understood before you try to make the case.

Get Your Ducks in a Row

Before making your pitch, understand what you need. Why is this important to the corporation? What, specifically, must be done to achieve service management excellence, and where do you need executive approval? The first thing you need is information, and we suggest having as much of the following at your disposal as possible. You can probably to gather much of it from efforts already underway in the company, and the act of collecting the data can help later as you work to build consensus among the various constituencies in the company.

1. *Where do we stand today?* Provide high-level measures of the company's strengths and weaknesses, benchmarked against world-class standards. These include customer satisfaction ratings, revenue per employee, and other productivity measures such as lines per employee, service quality performance, and service introduction lead times. Your information must be sufficient to make it clear that there is a problem (in the event that senior management doesn't yet realize it) or to make clear just how much more work remains to be done. This isn't always the easiest information to come by, but nothing motivates action like competitor comparisons—especially if your company is trailing. Market-research companies can provide information at a price, and some data is published in surveys by agencies and magazines.

   One of the best ways to gather this type of information is to exchange benchmarking data with companies that yours is not directly competitive with—perhaps a company that does business in a different part of the world. Such exchanges of data inevitably lead to discussions on approaches—even if it's how to get an investment case past a skeptical board. If you can't put together a comprehensive view, then pick one critical service (a real revenue-generator) and quote a competitor's tariffed prices and advertised service guarantees as a way to illustrate how tenuous your own company's position might be.

2. *What are our competitors doing?* Develop a high-level summary of your key competitors' strengths and weaknesses. If the company is intent on increasing global share, for example, learn who owns what and who is working with whom. It can be very effective to produce a chart showing just how the revenues are being shared, particularly if your company is a minor player with bigger aspirations. Before commissioning a new study or paying an outside analyst, check with the people who produce your company's strategic plan. They probably have the information and may even be able to give you charts, as reference points, that senior executives are already familiar and comfortable with.

   Being able to tie the discussion of service management excellence into concepts and strategies already accepted at the top is an excellent shortcut. Highlighting moves being made by competitors or at least well-respected industry players is another major motivator. Unfortunately, too much corporate strategy is of the "me too," variety, but it can and does motivate skeptical boards if companies they fear or admire are moving in a particular direction. Several major companies are beginning to make major investments in this area, involving billions of dollars of expenditure.

   Remember, turning the theory of service management excellence into a profitable reality is a multiyear activity, so project your position vis-à-vis the competition over several years. Prepare one view assuming the competitor goes for this approach and your company doesn't, and the other showing the results if you invest now. Don't assume the status quo—the industry will shift ground dramatically as competition and rising customer expectations rewrite today's notions of service levels and costs.

3. *What benefits can we expect?* Summarize, in order of priority of potential benefit, areas in which the company could benefit immediately from end-to-end process flow-through, including both internal and external interfaces. Keep the discussion at a high level—on the order of the NMF's Service Management Business Process Model—instead of delving into detailed flowcharts. Benefits should be clearly articulated in terms of unit-cost reductions, customer satisfaction improvements (which drive market share), and so on. If you are unable to develop a comprehensive view, it can be almost as effective to select a few good illustrations of how the company does something today (and associated cost, or delay, or dissatisfaction factors) and how the process could be improved. The example cited in an earlier chapter—of the company whose 60-day private-line installation interval included a mere 12 hours of real work—is a good model for this type of "sit up and take notice" illustration.

   To work, benefits must be clear and tangible. For example, we have repeatedly said that improving customer service will help improve profitability. How, specifically, in your company? A good example comes from

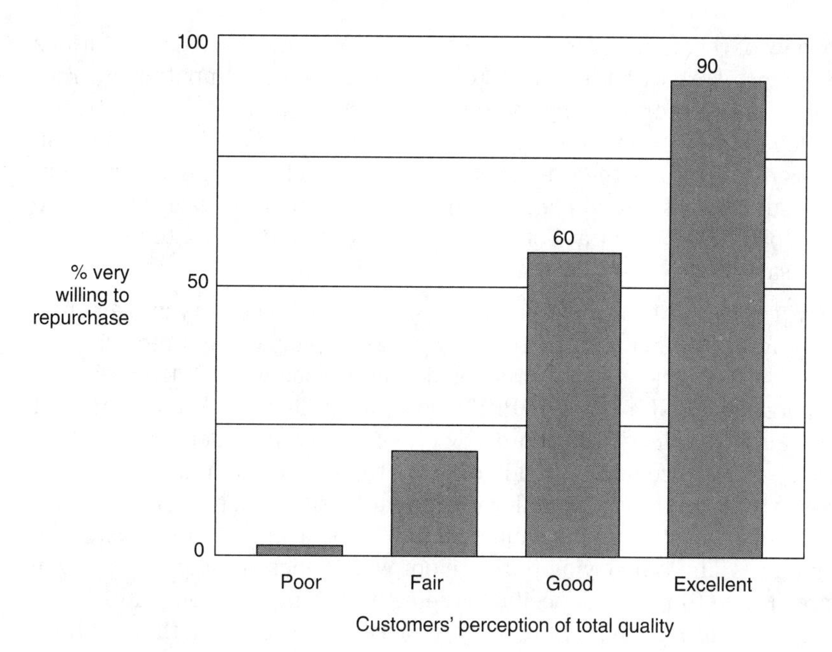

Figure 18.1 "Good" isn't good enough. *(Raymond Kordupleski and West C. Vogel, 1988.)*

AT&T, which found that just being good wasn't good enough to retain customers, as shown in Figure 18.1. In customer research, AT&T found that customers who rated their supplier as "excellent" were very loyal, but customers who merely rated AT&T as "good" could often be persuaded to switch to an alternative supplier.

Mapping this kind of analysis to specific services and analyzing how much revenue is at risk is much more persuasive than bland exhortations to improve.

4. *What have we done so far?* If your company has already reaped benefits from automating key interfaces (either internal or external), summarize the benefits that have accrued versus the front-end investment as well as the costs incurred. Try, if possible, to separate the one-time start-up costs (the costs to develop an infrastructure that is capable of supporting a common industry agreement) from the incremental costs of implementing specific company-to-company interfaces using that infrastructure. One U.S. provider of local services, GTE, having implemented a trouble ticket interface with two long distance companies, reported that the cost of building the first link was a seven-digit figure, while the incremental cost to implement the second link was in the five-digit range.
5. *What is the magnitude of the computing and process challenges?* If you have a specific process improvement in mind, present more than just

a view of the information that needs to be exchanged across the various systems. Work with the IT staff to understand which computing environments are represented by the different systems and what might be entailed in integrating them. Understand what some of the implementation options might be—the use of gateways, the development of a new application on a distributed platform, and so on—and make clear the need to approach the project at both the process level and the systems level at the same time.

6. *What is available to use?* Take stock of available industry implementation agreements that can be used by your company without requiring extensive technical work or prolonged negotiations with partners. This is where the OMNI*Point* and SPIRIT work can really help. Agreements that are already being implemented by other service providers or for which you can buy products should go to the top of the list. You want to stay away from producing a catalog of standards, since such a list will not be helpful in constructing an implementations plan, and you can waste a lot of money and time trying to figure out which ones are important. If you aren't able to relate a specific agreement to your own company's need, forget about it for now. Stick with things that can give you the best head start.
7. *What will it all cost?* You need to construct a reasonably complete but high-level (not too detailed) implementation plan in order to ensure credibility. It should state what needs to be invested, both in human- and financial-resource terms, and when (if a multiyear effort). It should estimate the length of time needed to see the desired results and summarize risks as well as contingencies that might be taken to mitigate them.

Developing the Five-Minute Message

Assume that you have only five minutes to make that sale to a senior manager. What are the points you must make? Here's our short list:

- Our business is making profits from supplying communications services. We remain competitive only if we manage those services extremely well—by delighting the customer, by continuously reducing our costs, and by improving our rate of innovation. Doing all three simultaneously is the way to sustain competitive advantage. It's called achieving service management excellence.
- We cannot sustain our current position by simply continuing to cut out fat. We will not achieve excellence without making significant changes in how we do business. Internally, we must reengineer our processes to gain speed and free up people to help us expand. Externally, we must put our-

selves in a position to do business globally (or regionally, or in a converged industry—choose whichever is most appropriate for your company), which means creating end-to-end process flow-through capability. To do this, we must achieve automated process flow-through, both between our internal systems and with our customers, partners, and suppliers.

- This can be done, although there are significant barriers to be overcome, and they are common to every other service provider as well. One barrier is that reaching interface agreements (agreeing on the "plugs and sockets") with other companies can be expensive and tedious. Another is that our systems infrastructure is outdated, and we need a computing environment that is only now beginning to become available.
- Solutions to these barriers are being developed by the industry, and we can take advantage of some answers right now, even as we combine forces with others to force more answers where barriers remain. There are ways to get started, and we don't need to make the entire investment at once. But we do need to make a commitment to this service management direction, because it won't happen overnight.
- If we fail to get started, we will lose valuable lead time to our competitors, including those we know of today and those that are just now entering the industry with a clean slate, which can design their processes and systems on a green field. This isn't an area in which we can respond "just in time." Building the strategies, competencies, and solutions takes considerable time. We must have the foresight to see we will need this in the future and start now.

These points are strengthened considerably if facts about the competitors' behavior can be sprinkled in liberally. If your company is becoming involved in international collaboration, as many are, point out that your competitors are providing end-to-end service management and offering global managed services across a variety of platforms right now. Delay could be fatal. Just how will you hand off that order or problem report to your new partner on the other side of the globe?

Depending on which efforts are already underway in your company, you may be able to shorten the message and cut right to the chase. A company with an extensive business process reengineering effort in place has probably already realized that it needs to streamline processes, but chances are very good it hasn't realized that to get end-to-end flow-through, it needs the support of others in the service delivery chain.

Blowing Away the Myths

If you are to be successful in getting a corporation to buy in to a major service management excellence program, you will not only have to point out the benefits, you will also have to dispel some of the myths and rumors about the subject. We've heard quite a few.

- *The value-added myth.* Once upon a time, cars were sold with heaters offered as an optional extra. Try selling even the most basic car without an array of creature comforts today and see how far you get! So, too, with communications services. Excellent customer service and an increasing level of service sophistication are becoming the standard in liberalized markets—not a value-added bolt-on to inadequate monopoly-style services. Service management isn't just an optional extra. Increasingly, it is a key ingredient of the services offering. Calculate your payback not in terms of value-added but in terms of retention or growth of market share. You might be lucky and extract a premium for good service management, but treat it as a bonus in your calculations—not as a sustainable revenue source. Profits come from leveraging service revenues and cutting costs.
- *It's all too difficult.* Getting to service management excellence is a long haul with some significant technical and operational challenges along the way. Most companies hit problems in internal organization and discipline. A myth that can grow up is that a cross-company activity like this one is just too hard. Even worse is the "we tried it and it didn't work" retort. The plain fact is that sooner or later, every service provider (at least those who stay in business) needs to go down this path. If the current management team can't manage it, then someone who can will.
- *The standards aren't stable.* This one is a favorite chestnut of many suppliers. Sales training across the industry seems pretty uniform, because virtually every supplier (of equipment or service) has trotted out this myth at one time or another. The truth is that the basic building blocks of the management standards we mentioned in Part 3 are now very stable. They've been washed, polished, profiled, tested, debugged, and implemented time and again. Some of the uses of these standards and consequent object libraries may be new, requiring work between customers and suppliers, but that's like discussing the style of your kitchen cupboards—not whether the foundation and walls of the house are stable. Tell your suppliers—very simply but firmly—to play another record if they try this one!

Choosing Your Approach: Speed or Safety?

Any business case must be related to specific plans or activities, but most service providers are in the unenviable position of having no end of opportunities to improve their service management processes. Timing, as with comedy, is everything. So what is the best way to get started?

Point solutions: The hare vs. the tortoise

A "point solution" (sometimes called the stovepipe approach) is one that has a specific purpose and that is implemented in an isolated way—that is, not as a part of an overall strategy. An example is implementing an automated trouble ticket exchange with another provider that is proprietary and that only applies to that one business interface. The advantage is speed and a more contained approval process—if few people are affected, then getting buy-in isn't too hard, and early successes can be very useful in bringing home real and tangible results from a limited investment. However, you need to know how to move the point solution into a more comprehensive approach as you get more successful.

Without a migration strategy, the disadvantage of point solutions is that they don't fit within the mainstream of systems or processes. They will need to be separately maintained for as long as they exist, and they may not help the company move toward a more general goal of continuous service improvement. Also, although initial investment may be smaller than that to establish a comprehensive service management platform, multiple single investments over time are likely to cost more in total than if the company were to commit to an overall strategy and invest in infrastructure that can serve the full range of service management needs over time.

Standards have a terrible reputation, for good reason. They are slow to materialize and often do not deliver what's needed. But when an industry has gone to the trouble to make standards useful as well as easy to use, ignoring them may be unwise. If a nonstandard approach is adopted, it means that your company will have to address both the systems-level and process-level integration issues, which are considerable. Here is a brief summary of what is involved if a proprietary approach is undertaken:

Systems-level integration. Using a nonstandard approach means that common software platforms will not be capable of supporting your applications, except through customization. To what extent do you want to spend scarce resources developing (or paying to have developed) a unique platform, which will then need to be uniquely maintained throughout the life of the systems that it supports? What business benefits can be cited from doing things differently at this level? Your development team may be happy, but in our view, it's a losing business proposition.

Process-level integration. The service delivery chain may be long, and the use of nonstandard information-exchange agreements must be accepted by every single player along the chain if end-to-end process automation is ever to be achieved. How many equipment suppliers are willing to support a proprietary interface (other than their own)? How many other partnering service providers are willing to pay for custom development to be able to interface with your company? How many years are you willing to spend to devise individual process flow-through agreements that are consistent with your (nonstandard) approach? And how many customers are willing to support a nonstandard interface or make room for your proprietary system in addition to the one they use for everything else?

A comprehensive plan: Smart but slower

The most successful companies will be those that take a more comprehensive approach to service management, developing a set of policies that govern procurement and development and striving for consistency as each service management improvement is undertaken. Because of the number of constituencies involved, this is a long and difficult task to accomplish, calling for a significant degree of cross-company cooperation and agreement. However, once the decisions are made and the initial investment is past, additional service management improvements can be made very quickly and at a small incremental cost.

The first companies to face the need for a service management strategy invested huge sums of money, not only in developing their strategies but in determining what type of infrastructure support would be required and then driving suppliers to deliver that support. Fast-follower companies—those coming along behind the pioneers—have a tremendous advantage, because the early leaders insisted on establishing their strategies based on standards. Although the time to develop a strong, company-wide policy toward service management process integration is still considerable because of the many constituencies that must be convinced, at least now there are clear answers about which direction to go, and there is growing product support to make the job easier.

Standards-based point solutions: One step at a time

Another, perhaps more practical or pragmatic, approach is to target a specific service or process for improvement and to define that project as phase one of a multiphase, comprehensive plan. In other words, under this approach you would actually develop a comprehensive plan and business case to guide your approach, but funding requests would be made one at a time, with each one building on the former and each one demonstrating benefit to help "sell" the next one. This has the advantage of taking the skeptics with you (provided each implementation is a success!).

Companies only now beginning to address process-level integration have the luxury of adopting this third approach, because earlier pioneers have paved the way. Take advantage of this where you can. The main downside of this approach is, again, time frame. Your competitive situation might not allow you the time to proceed along this step-by-step pathway. Nevertheless, it's worth starting on a pilot basis, even if you progress to a more widespread, parallel approach later on.

Chapter

19

Building an Individual Project Case

Let's assume that you will be following the third approach proposed in Chapter 18—presenting a business case for a specific project based around a particular service or process, within a well articulated overall plan and framework. Selecting the project is an important step, since some projects are more difficult to prove in than others. Several factors should help guide your priorities:

- Where is the company "bleeding" from excessive costs or service quality that doesn't measure up?
- Where have industry agreements already been reached that might be supported by available technology or at least willing partners and a strong support group?
- What specific service measure are you on the hook to improve?

This last point will likely guide your first moves, but don't forget the other two. An early "win" that makes everyone on the team a star can set the pace for the next hurdle, and you should use every tool to your advantage. If you can make a case for improvement in an area where other companies are active, and your company has an opportunity to learn from their efforts, so much the better. Why take on a problem that no one else has tackled unless it happens to be the one thing you are being held accountable to fix?

Make Use of Tried-and-True Quality Principles

Most companies have some sort of total quality program in place, and the principles of quality management offer a great deal when determining where to concentrate your priorities toward service management excellence. There are volumes and volumes written on this subject, but here is a brief summary of some of the key principles to keep in mind—principles we both have employed successfully before charting service management improvements.

Reduce the number of steps in the process

Champy and Hammer point out that breaking work down into very small steps was a positive thing to do in the early days of the industrial revolution. Each job could be carried out by a cheap, low-skilled worker. But the more steps, the more likely the process will suffer from failure. People make mistakes, and so do systems. Assuming that each step in a process attains an average of 99.5% accuracy and on-time performance, a process with eight steps delivers a cumulative average of only 96% expected results, as shown in Figure 19.1.

An average of 96% might sound pretty good, but in an environment in which the process is in use 24 hours a day, 365 days a year, 96% success means that the process will be broken for more than 14 full days per year! A process with only four steps, in which each step achieves 99.5% success, yields a total process success rate of 98%. By reducing the number of steps, significant improvement can be realized, even with no underlying improvement in each step. Getting to process automation means eliminating people from the process wherever possible, and, since sophisticated systems now form the process chain, the number of steps can be successfully reduced in a reengineered environment.

Improve the quality results of each step in the process

Every step in a process involves a customer-supplier relationship, with "suppliers" providing needed input, and "customers" receiving the output. In a typical service order process, for example, the service order entry center looks to the sales department (or the customer) as the "supplier" of initial information and includes as "customers" the operations department, the customer engineering center, and the billing system, each of which might have different needs. In today's world, that order entry center is

.995	×	.995	×	.995	×	.995	×	.995	×	.995	×	.995	×	.995	=.96

Figure 19.1 Reducing the steps in a process generally yields improvement in overall quality.

likely to generate a single output to all three "customers." Each of these departments, in turn, is required to sift through the service order to find the information it cares about, and each might need to translate the information into terminology that matches its own systems or methods.

The use of common terms can help avoid error-prone translations. It is not uncommon today for a service provider's many systems to use a dozen or more different formats to represent the customer's name. The mere act of translating the same basic information from one step in the process to the next creates errors while adding no value. By using a single format for each piece of information, errors are eliminated and translation time is saved.

Reduce manual data entry

Another quality improvement involves making better use of information that already exists somewhere in the company. Instead of reentering data at multiple steps (and making errors in the process), making better use of existing data can result in significant improvement. For example, when generating a change to a customer's existing service, it is much smarter to access the customer's exact name from existing account information than to reenter the information from scratch.

In general, any reduction in manual intervention improves quality dramatically. Certain things can only be done well by humans, such as "sending a smile" to an irate customer with a tone of voice that says "I care" or making planning decisions based on performance trends. Transcribing and formatting data is better done by computers. It's very much like the difference between the days before the PC—when a letter had to be typed and completely retyped each time a change was made (usually adding new typos each time old ones were fixed)—and today's office environment. Think about how most people work these days, and how the reduction in manual intervention has improved productivity. Rather than writing out letters longhand, then giving them to a clerk to be typed, then correcting and reediting, and so on, most people compose letters and presentations electronically, making their own changes and edits as they go. In far less time than it used to take to deliver a plain, typed letter, technology makes it a simple matter to produce a beautifully formatted, spell-checked, illustrated briefing paper.

Employ the Pareto Principle

The Pareto Principle simply states that not all problems are of equal importance or represent equal opportunities to reap gains from process improvement. Some processes probably work quite well just as they are. Some don't really lend themselves to automation, perhaps because the activities are not terribly repetitive. An example is the process of developing a new service. Although the service development manager goes through the same steps each time a new service or feature is developed (e.g., creating a technical service description, developing a service operations and testing plan, performing operational readiness tests, and so on) each service is different from the rest. Templates for "how to deliver a service" can be developed to improve consistency of output, but the steps can't be mechanized in the true sense of the word, and a lot of time and money could be wasted trying. Automation efforts have the greatest impact in areas where problems are recurring and where they are most damaging to overall service delivery performance. As a general rule, 80% of a problem can be attributed to 20% of the process, so fixing the right 20% yields big results.

Use a Service Quality Methodology to Identify Areas of Weakness

Launching a major activity to reengineer service management is, as we have said, a major undertaking that can easily become discredited if it is not focused and directed. Once large numbers of people start to be involved, it is all too easy for the activity to lose its way and become dissipated. Identifying and focusing on the areas that will bring the highest payback in the shortest time frame is essential to gathering and maintaining service-level support for the program. Thus a strong methodology to quantify and identify targets is essential.

An approach we have found useful is the SERVQUAL methodology developed by Zeithaml, Parasuraman, and Berry in their book *Delivering Quality Service.*[1] This model uses a gap-analysis technique to track five service attributes: service reliability, empathy with customers, service level assurance, customer responsiveness, and tangibles such as premises, vehicles, and service personnel appearance. Although the model is generic to service industries, we believe that a rigorous approach such as SERVQUAL for delivering service management excellence is essential for aspiring lean communications service providers.

In essence, the method focuses on answering the question "How can we continuously reduce the gap between what service customers expect and what they perceive they are getting?" Ideally, the lean service provider aspires to exceed expectations and be the market pace-setter. The key to closing this gap (shown as Gap 5 in Figure 19.2) is to focus on closing four other important gaps and keeping them closed.

Let's look at each of these gaps in more detail.

- Gap 1: The gap between the service levels that customers expect and what the service provider perceives they expect. For the provider that

---

[1]Adapted/reprinted with the permission of The Free Press, an imprint of Simon & Schuster, from *Delivering Quality Service: Balancing Customer Perceptions and Expectations*, by Valerie A. Zeithaml, A. Parasuraman, and Leonard L. Berry. Copyright 1990, The Free Press.

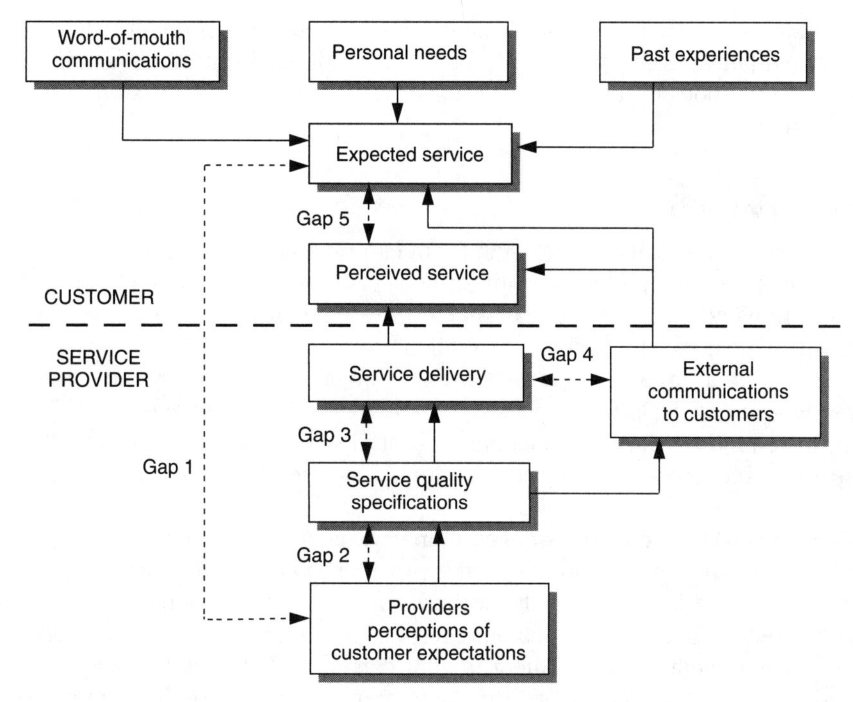

Figure 19.2 Customer service model. *(Zeithaml, Berry, and Parasuraman)*

has yet to grow the skills of marketing and customer-needs analysis, this can be a difficult area to manage. Much of the telecom industry has an arrogance borne out of many decades of telling customers what they can have, rather than being attuned to listening to what customers want.

- Gap 2: The gap between the service provider's perception of customer expectations and service specifications. These specifications drive (or should drive) process and systems development activities and all of the factors we have discussed, such as process flow-through. If these are developed in a vacuum and do not take into account what people who speak with customers believe to be the customers' requirements, then getting to excellence is significantly at risk.
- Gap 3: The gap between what is specified and what is delivered. This is the classic process and systems dilemma—how to design and implement services that meet specified performance for cost, quality, and time scale—our three-legged stool problem in sharp focus. Inevitably, because these objectives are opposing, tradeoffs are made during the development phase, making it very difficult to close this gap.
- Gap 4: The gap between what is delivered and what customers are told they should expect. Earlier in this book, we described a trial of online

customer management that, although it didn't change the underlying service quality, radically shifted the customer's perceptions of service quality to the bottom-line benefit of the service provider that invested in the change.

Closing the gaps

Measuring these gaps and closing them can be a very powerful way of focusing on process and system investment priorities. This is essential if the sort of business results that can fuel a wide-scale backing of investment in the goal of service management excellence are to be achieved.

If this is an area of your responsibility or one that interests you, then reading the book *Delivering Quality Service* is highly recommended. For an introduction, the key concepts are outlined here, modified to suit the service excellence ideals of the communications service provider.

Closing Gap 1: Finding out what customers expect. Gap 1, the gap between customer expectations and the service provider's perception of these expectations, is a key issue, especially for those companies that have not developed competencies in measuring customer needs. To determine whether this might be an area needing improvement in your company, Zeithaml, Parasuraman, and Berry suggest looking at three factors: marketing research orientation, upward communication (between customer-contact personnel and senior managers), and levels of management (implying that too many levels can inhibit the flow of useful information).

In essence, these factors have to do with how a service provider is structured (and how it functions) to guarantee a consistent approach to learning about customer expectations. It is well and good for a single product manager to do a good job conducting focus-group interviews and market research, but if the corporation does not foster regular behavior in this area, then service quality will be inconsistent at best. Similarly, some managers may do an outstanding job of encouraging communication between their lower-level customer-contact personnel and senior managers, while others may keep lines rigid. Service quality results are likely to vary widely between two such groups in terms of their ability to understand what their customers expect.

Learning what customers expect can be among the most difficult challenges for a newly liberated service provider. Aldebert Wiersinga, manager of the OSS Competence Center of Cap Volmac, a large systems integration company and part of the Cap Gemini Sogeti group, notes that it is difficult for many service providers to invite criticism from customers. But, as he explains, "If customers stop complaining, the operator stops competing."

Unless you are in control of the factors surrounding Gap 1, all of your efforts to reach service management excellence are based on internal perceptions and good ideas. If you don't really know what customers want, your efforts in reengineering and systems integration stand the chance of being misdirected.

Closing Gap 2: Setting the right service standards. Translating customer needs into the service specifications that drive process and systems development is essential if the right investment is to be made in the right area at the right time. It is hard to believe, but too often service specifications are written independent of perceived customer needs. This occurs when a company is not focused on service excellence and lacks the commitment to develop such a culture. According to Zeithaml, Parasuraman, and Berry, this shows up in factors such as the level of commitment to service excellence (as demonstrated by company actions); the belief (by those doing the job) that it is possible to meet stated customer expectations with the processes, methods, systems, skills, and people available; and the demonstrated setting of personal performance goals that are directly tied to customer feedback.

This set of factors has some very interesting nuances for communications service providers. For one thing, the development of new services must constantly be weighed against improving existing ones. Organizationally, if the management of new service development is widely separated from the management of existing services, the battle for funding and recognition can potentially leave the customers' needs and service quality excellence a distant third in consideration. Conversely, if senior managers have product management, development, and operational responsibilities for the entire life cycle (inception through withdrawal), decisions are generally made with the customers' best interests at heart, and market research efforts are able to discern whether customers are more interested in added features or improved quality.

Another key point can also be made here. Successful service development activities must start with a well-defined service description that tells how the service or feature is to be used, technical quality standards and cost objectives that must be met, and operational customer service levels that must be supported. It is not enough to simply request the ability to give customers control over the configuration of a service, for example. Information must also be included with regard to the number of customers who must be supported, system response times that must be met, service guarantees to be included in the SLA (and how they will be managed), ordering intervals, billing accuracy, bill format options, and so on. Without this level of service definition, the "hooks" will almost certainly not be built into a new service to enable proper life-cycle management.

Closing Gap 3: Delivering on stated service performance objectives. Gap 3 is the heartland of our book. Close this gap—the gap between service specifications and actual service delivered—and you will really be making progress. This is the management of the processes, systems, people, and interfaces that comprise our three-legged stool. With Gaps 1 and 2 closed, Gap 3 can be viably closed only through system and process integration.

In an environment in which people are being removed to shed operating costs, they must be replaced by technology, or service levels will not meet desired results. This balancing act between people and technology to deliver service excellence is crucial, especially across the boundaries between processes and between organizations involved in the process delivery chain.

Zeithaml, Parasuraman, and Berry point to a great many factors that influence success in closing this gap, including role definition (is everyone playing from the same song sheet?), role conflict (can people get the job done without always having to jump through hoops?), employee-job fit, appropriate use of technology (do management systems improve flow-through or inhibit it?), control systems (is good performance rewarded?), perceived control (are people able to make necessary decisions?), and teamwork.

We would add another factor, bottlenecks, with the notion that it should be possible to see where costs or delays cluster, where process flow-through improvements can help the most, and where inflexible systems lead to protracted service introduction schedules. There are tools available to help service providers model their processes, including simulation tools that identify bottlenecks and their causes. But don't let yourself get carried away by the tools and forget what you're trying to determine. As Cap Volmac's Wiersinga cautions, "Keep it simple—low-tech, high-touch" in order to get the best insight into how internal processes might be improved to raise customer satisfaction. Tools are only an aid and can't begin to substitute for working with the people involved to get a true appreciation for where problems exist.

As for avoiding problems in the first place, service providers are under tremendous pressure to deliver new services faster than ever before, and often they forget about many of these factors. Somehow, in the planning stages of a new service, the wonderful flexibility of the human is taken for granted, making for a multitude of sins in the service design and development phases. However, using people as bandages and string to hold a process together is a very short-term option. While it might be acceptable to get a service to market quickly, costs will kill profitability in a competitive market. The aim is to achieve high levels of process flow-through without human intervention.

It can be very difficult, as an operational person, to block the introduction of a new service, but the alternative—managing something without the right structure or tools—is worse by far. A technique that is fairly widely used and is essential to success is conducting operational readiness tests (i.e., testing all operational processes) as well as technical service tests before permitting a new service to be sold. For services that involve a fair degree of tailoring for each customer, such as a global virtual private-network service, this operational testing phase may be called a beta trial, involving a small number of actual customers who are given the service to try during a defined period of time.

For services intended for mass sale, in which the operational processes need to be especially reliable and well-linked, the operational readiness test must simulate actual behavior of a fairly large number of customers. One way to do this is for members of the service development team to act as customers, placing orders for the new service and exercising the actual processes, methods, and systems that would be used during service delivery. This type of test needs to be of sufficient duration to test the complete cycle—including billing, problem handling, and so on—and many problems are usually discovered and fixed during this period.

It is interesting to note that in evaluating this gap—the difference between services as specified and services as delivered—there are many factors to consider that go beyond the technological integration of processes and systems. It's a good lesson to keep in mind, because no matter how effectively processes are designed or linkages are created, if attempts to improve process flow-through do not also consider the human aspects listed here, success will remain elusive.

Closing Gap 4: Resisting the urge to overpromise. This gap is a major plague for new and existing service providers alike. Sales and marketing teams, hungry to make inroads into an incumbent's market share or defend an existing customer base, can often promise more than the process and systems infrastructure is capable of delivering. How often has a senior manager, when pinned down in a customer or shareholder meeting, promised to deliver a certain level of performance to the horror of the operational team that has to implement it? (One of us remembers manipulating user-group meeting schedules so that a senior VP who was notorious for making impossible promises could not attend.)

Yet perception is reality. Companies that earn a "halo" effect by providing consistently excellent service are more likely to be forgiven by their customers for an occasional lapse. Those companies that usually miss the mark, however, are condemned even when they really are making an effort.

A few years ago, one major global provider found that its pay phone business was losing money. It made a decision to cut back on maintenance of public pay phones, with a consequent drop in service quality. Although the impact of the cut was negligible to the bottom line, the company found that by letting its public pay phones slip, customers' perception of their other (profitable) services began to suffer. Said one large customer, in rejecting a bid from this company to provide a major global value-added network, "If you can't even fix the pay phones, how can I believe that you can implement and manage a state-of-the-art corporate network?"

The authors of *Delivering Quality Service* point to two key factors that relate to this gap: horizontal communications (particularly the extent to which the knowledge of contact personnel is used as input to advertising strategies or the forewarning of these personnel that a major new campaign is about to be launched) and propensity to overpromise.

Think about how you feel, as a customer, when you call a company in response to an ad, only to hear a long silence at the other end! The provision of ISDN service seems to be a favorite "black hole" for U.S. service providers at the moment. Normal contact personnel seem never to have heard of it, or if they have, they direct the customer to a special number that is usually connected to an answering machine. Saying you provide a certain service is one thing, but if the people who must deal with customers on a regular basis are not prepared to answer even the most basic questions or to direct a call quickly to someone who can help, the customer is not well-served.

Closing Gap 5: Getting to excellence. All of the gaps outlined in *Delivering Quality Service* must be examined and closed if you are to achieve excellence: meeting or beating the customers' requirements for price, quality, and timeliness.

With so many factors to consider, there is a real danger of getting engrossed in a single facet of service delivery, such as cross-constituency linkages, and losing perspective of the business objectives that caused you to consider instituting those automated links in the first place. What you need is a process to improve your processes, a way to continuously monitor the gaps, take corrective action, and remeasure. The simple flowchart shown in Figure 19.3 is a good aid in keeping the high-level view. Within each of the major areas of investigation (shown as the shaded boxes in Figure 19.3), you can apply the specific factors cited earlier in association with each gap identified in the SERVQUAL method.

The point in examining a fairly generic service quality process such as SERVQUAL is to remember that achieving service management excellence is a many-faceted undertaking, involving issues of policy, method, and technology. Applying extraordinary efforts to forge intercompany agreements without also dealing with the many internal issues of process improvement will yield marginal results. Although your specific role in service management improvement may pertain to only one or a few of the many factors cited in SERVQUAL, as a participant you should understand how your efforts fit into the broader spectrum of improving service quality. If the link is not clear, start asking questions!

If You Can't Measure It, You Can't Manage It

The SERVQUAL methodology relies on a gap-analysis approach to improving service management. But gaps become obvious only through measure-

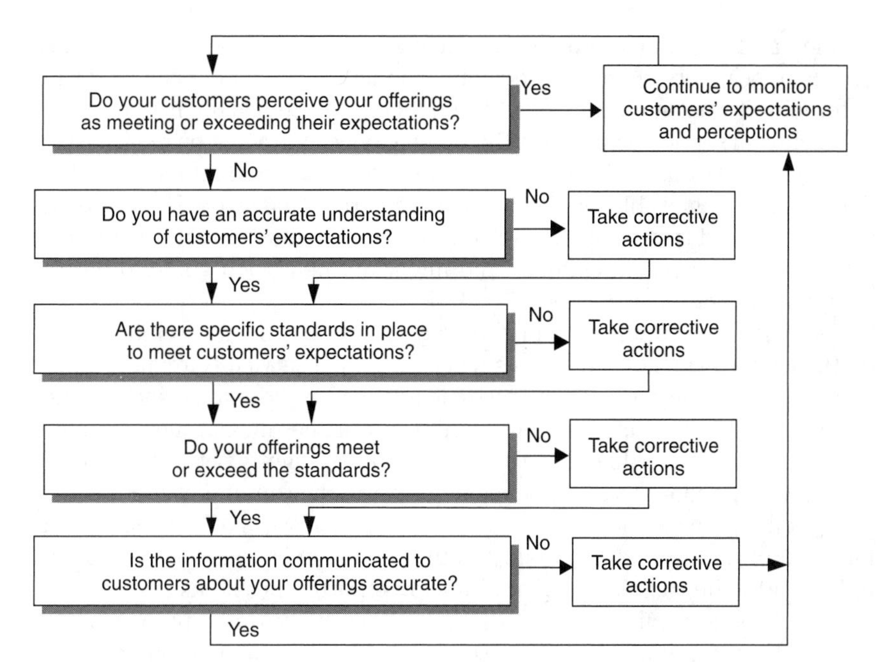

Figure 19.3 Process model for continuous improvement. *(Zeithaml, Berry and Parasuraman)*

ment, and perceived improvements that can't be measured may just be wishful thinking. Evolving a set of meaningful metrics is one of the more difficult challenges, but it is key to achieving excellence. Nothing succeeds in focusing the behavior of an organization like clearly visible and measurable targets—especially if they are linked to rewards.

This is pretty basic management philosophy, but it is no less important because of its simplicity. Unless you are able to articulate clearly where you are, and where you aspire to be, a service excellence program will take the same path as many ill-managed "quality improvement programs"—becoming nothing more than exhortation for things to somehow be better than they are today.

The overall goal of corporations striving for "lean operator" status is to increase return to shareholders. Contributing to this overarching goal, the operational service management goals are to reduce cost, improve customer service, and shorten time to market—the three-legged stool. Within the context of these three operational goals, how can metrics be established that help service providers make positive progress?

Measuring service quality

Measuring service quality in practice demands that you build probes into your customer service delivery channels that can drill down into the organization and pinpoint where improvements are being made (or not, as the case may be). Service level agreements should certainly be a key factor in determining what measures need to be taken. These usually include both the technical service parameters that are of concern to a customer (i.e., service availability) as well as some of the softer factors, like meeting due dates, correcting problems within a defined period, and so on. But it should be remembered that existing service level agreements may reflect the service provider's capabilities today and might not be a good guide to future competitiveness.

For example, many service providers give their customers technical data on the performance of the various piece-parts of a service (circuits, equipment, etc.), or they provide average measures of performance such as the "mean time between failures" on the network or the "mean time to repair" a network problem. While these measures may be slightly better than nothing, service providers should be moving toward measures that more accurately reflect the customer's experience with the services offered, such as a measure of end-to-end availability from the customer's perspective. Similarly, in addition to measuring how quickly reported problems are cleared, service providers might want to look at how many times problems are cleared by the initial contact person while the customer is on the line, which gives a good indication of how well the processes and systems are supporting the employees in their desire to provide good service.

Results should be tabulated depending on how the company is (or wishes to be) organized—by service type, by functional group, by geographical region—so that it is possible for a defined group of people to have a positive effect on improving service to their customers. Team awards that encompass the entire company may initially help to encourage people to be concerned about company profits, but they have little or no effect on improving operational performance.

A word of caution is in order, however. Avoid the temptation to institute strict targets before it is possible to capture the needed data or before you have been able to confirm that the measures you are striving for will truly produce the desired effect. For example, if one aspect of providing good customer service is handling calls promptly and quickly, and you choose to measure call length, you might find that contact personnel begin to talk too fast to be understood or brush customers off before their problems have been solved.

Measuring time to market

Product introduction cycles vary widely depending on the service, and there is a vast difference between introducing a totally new service (which is dependent on new technology being developed and implemented) and introducing a new feature on an existing service. Technology limitations are being overcome by moving intelligence out of the switches, by creating modular switching and signaling capabilities that can be invoked in different combination, and by moving toward standard ways of manipulating existing equipment through changes to the element management systems.

Measuring the effects of these changes can be done by looking at the total time to design a service; the time needed to complete any equipment development or changes; the time needed to design and then develop (or modify) all of the operational support systems, including both the network-facing and customer-facing systems; the time needed to test the technical capabilities; the time needed to test the operational processes and systems; and so on. Although each service or feature introduction is different, most service providers manage enough projects to be able to develop appropriate targets for different types of projects. It should be possible to develop several categories of service development project and to establish targets for each category. If you get specific rather than apply a generalized plea to "go faster," employees have an opportunity to succeed and improve on results, rather than becoming frustrated.

These time factors, of course, must be tempered by quality factors. If a service is introduced quickly but can't be managed well, speed should not be rewarded. If the stated delivery interval of a new service, for example, is five business days, then the service development team should be held accountable if that target cannot be met, since it means that either the systems or processes did not meet the requirement or that training of personnel was inadequate, and employees were not properly prepared for the service launch.

Proxy targets

Cycle-time targets for product introductions can be articulated and measured without too much difficulty, provided that the product development process follows some sort of established phased review process that ensures a high-quality output and marks clear start and end dates within the process.

But what about cost reduction? What can be measured, and how is it possible to be certain that reductions in operating costs are due to service management excellence programs and not to other factors, such as expense controls, astute pay negotiations, or hard-nosed procurements?

Setting targets relating to cost reduction of processes can lead to distortions. People find creative ways to massage figures in the short term, such as reallocating overheads to different expense lines to give the appearance of a cost reduction. Somehow, organizations are at their most innovative when trying to make the bottom line look rosy!

To overcome these pitfalls, it is often useful to set targets that are a proxy for the real target of cost reduction. For example, managing cost in an airline is difficult to achieve for product managers where myriad cost factors are outside the direct control of the people responsible for route profitability. They overcome this problem by managing proxy measures such as load factors. The higher the load factor, the fuller the plane, and the more profitable the route. Load factors are maximized by either selling more tickets (revenue maximization) or reducing the number of available seat-miles by flying fewer or smaller planes (cost minimization). In any event, the measure reflects the overall desire of the airline to make routes more profitable and so encourages the right kind of behavior from the product manager.

Finding ways to measure cost reduction in the telecom industry—particularly cost improvements associated with reengineering and automating process flow-through—is another place where proxy metrics might be used.

At the heart of any reengineering activity is the belief that the replacement of people-based bottlenecks, queues, and inefficiencies with integrated service management systems will result in cheaper and more effective ways of serving the customer. If the goal for a certain process is to achieve "one-touch" or "no-touch" process flow-through, then one interesting indicator of forward progress might be the percentage of the process that is automated. By establishing reasonable "stretch" targets that permit an organization to automate a portion of a process as a first step and then add to that over time, progress can be charted. One way to measure this might be to count the total number of manual interventions in a process and set specific targets for reducing that number in stages over time.

Another measure might be to relate the number of employees to the revenues generated by services, distinguishing as much as possible by service type. Here, as in the airline case, lowering this ratio can be done by increasing revenues or by taking out cost (automating), and both are generally positive behaviors to encourage. However, as we pointed out in the first pages of this book, it is also possible to gain at least short-term improvements in cost by cutting back on quality of service, just as it is possible to bring in new revenues by overpromising what can be delivered.

The 3LS measure

Individual measurements of service quality, costs, and time to market must be made, because all three are important contributors to producing revenues and profits, and quite often different groups may be key to realizing each goal. But even more important is making certain that you are not sacrificing one goal to make gains in another. For that reason, individual measures should only be used as indicators, while the true measure of improvement is a combination of the three. Call it the 3LS measure, as a way to remember the three-legged stool!

An overly simplistic 3LS measure, then, might involve creating indices for each of the three main categories of measurement and then adding them together in some way (again, by service type or other grouping) to produce an overall service management excellence index. To focus on the issues that have been raised in this book, it is important to separate the operational measures from overall revenue or profit measures, since many other factors will be included in the overall corporate figures.

The task of establishing metrics is a significant challenge, and if simple measures can be found, these are often the best to use. The important point is that without metrics, you will not be able to determine whether you are getting value from your automation projects, nor will you know whether you are making progress as fast as you need to.

The People Part

There is a real tendency to think that automation is just about technology, but that is certainly not the case. As most people find, the technology is usually the easy part. Much harder is knowing where to start, how to set priorities, and how to bring the many different, often divergent, groups together to support change—in other words, building a learning organization.

Service improvements cannot be accomplished in a vacuum. Service management excellence isn't something that "someone in operations" does. It involves the entire organization and its extended value chain of suppliers, cooperating providers, and customers. As such, it demands attention in the boardroom and requires the drive and energy of very senior managers who can cut across the fiefdoms. It requires the understanding and flexible cooperation of suppliers that, if they get it right, stand to make significant sales to aspiring lean service providers. Finally, it requires an understanding across the industry of the issues, a common language and approach, and a knowledge of when to cooperate and when to compete.

Many different constituencies need to be involved when a service management improvement is contemplated. The technical teams need direction from marketing. The operations teams need to involve systems development, and so on. In the next chapter, we look at these constituencies more closely.

Chapter

20

Working with Constituencies to Effect Service Improvements

Five groups of companies share an interest in service management. The first group of companies is comprised of the service providers—an entire range from traditional monopoly operators through newly formed start-ups. The second group consists of telecommunications equipment suppliers, again of many types, from huge corporations covering a wide spectrum of technologies through highly focused, young companies specializing in a new, red-hot technology. The third group is made up of computer suppliers covering a range of sectors—hardware companies, software companies, applications providers, specialized management system and applications vendors—all active and growing quickly as the needs of the service provider industry unlock huge procurement budgets. Fourth are enterprise customers—large commercial corporations that run major internal private network. Last, but certainly not least, are the growing band of content providers—organizations that provide anything from recorded weather reports to online databases and videos on demand. Within each of these five major groupings are myriad internal constituencies, each of which may sometimes act independently of the others, sending mixed signals and interfering with attempts to implement a cohesive strategy.

Depending on the type of service management improvement you intend to achieve, you need to deal with several internal constituencies as well as multiple external constituencies. The key groups, and their perspectives, are outlined in the next several pages.

The Many Constituencies within a Service Provider

Service providers often employ tens or even hundreds of thousands of people. Although each company is organized differently, in general, a number of major constituencies can either help or hinder the pathway to service management excellence through automated process flow-through.

Operational groups

Operational groups include both the people who manage the customer interface and the people who manage the network. In some companies, there is a defined split between those involved in the "business office processes" of sales, ordering, and billing and those involved in the "network operations processes" of problem handling, service configuration, and network management. In other companies, a single customer operations group brings together all of the functions that are related in some way to delivering individual customers' services, while a network operations group manages the infrastructure, including network design and installation activities as well as real-time network monitoring and maintenance.

The operations group is most directly affected by any changes to processes and operational support systems and also has the best ideas for improvement. This group has a perspective based around customer satisfaction and process efficiency, combined with significant expertise in networking. However, operations groups are often rather inward-facing and might not be aware of industry efforts to address service management improvements that need to extend beyond their companies' borders. Also, although well-versed in issues as they relate to existing operational support systems, these groups may have limited knowledge of complex, cross-domain computing integration issues.

These groups are probably not interested in acquiring general knowledge about what's happening elsewhere in the industry. However, where specific industry agreements exist in an area that matches the target of the proposed service improvement, the operations departments should be made aware of those agreements in detail. Even if the proposed improvement is seen by these groups as being an internal issue, point out the advantages of implementing the change in an industry-accepted way so that links can be forged with external companies at a future date without requiring further change.

The procurement department

The second constituency within a service provider is the procurement department. The average procurement budget of a major service provider runs into billions of dollars annually, so these departments are often very large, and they are often tightly connected to network and systems architects and other planners. Policies that drive procurement actions generally include conformance to international or regional standards, although typically these are referred to rather generically, giving suppliers wide latitude in interpreting the standards. A popular requirement stated in service provider procurement specifications for telecom equipment is to "comply with TMN," which, although useful as a directional statement, has very little meaning when buying a management system or managed equipment. As discussed earlier, TMN is a framework, not an implementation agreement.

The perspective from the procurement group is to try to maximize the value of supply contracts, and the emphasis is clearly on the capabilities of the equipment rather than how that equipment is managed. Often, short-term views predominate, and mistakes are made in awarding contracts based on initial purchase price rather than looking at the life-cycle costs of bringing a new technology into service. This is where integration issues at the network and service management levels are crucial. A technology that cannot be integrated to achieve automated process flow-through demands more people to act in that role, forcing up the lifetime costs of such a procurement.

This constituency needs to understand the issues of automated service management and the relationship between its actions and the company's future ability to compete to appropriately deal with suppliers to gain the desired result. In addition, the members of this constituency need to be schooled in the advantages of procuring against specific implementation agreements as opposed to simply giving suppliers a long list of standards or requesting "compliance to TMN." Again, the dictum "if you can't measure it, you can't manage it" applies—how can you structure a contract with payment against deliverables unless the implementation details are highly specific? Without this information, both sides could be in dispute for a long time. Writing these implementation agreements takes a long time and costs a lot of money—why bother when much of the hard work has been done in the OMNI*Point* Solution Sets outlined earlier?

R&D and IT groups

The third major service provider constituency is the research and development organization. This group often has the skills base to understand the advanced technologies that need to be deployed to achieve automated process flow-through. The problem is that although the members might be highly skilled in this area, they sometimes lack practical experience of network operations and process design and might let themselves be guided by technology rather than service drivers.

Closely linked to this constituency is the information systems (IS) or information technology (IT) group. This group is often responsible for operating the major systems that have traditionally supported the customer care service management activities: the major billing systems, customer administration systems, order handling, procurement systems, and so on (as opposed to the network operations support systems, which are usually within the domain of network operations). Many of these systems are traditional in design, often based on mainframe technology and generally stand-alone, i.e., not integrated into the process flow but "fed and watered" by large numbers of people involved in delivering a particular service process.

This constituency sometimes lacks the necessary experience of systems integration, object orientation, or client/server distributed systems that are part of the technologies being deployed to achieve automated process flow-through. As this group generally is also responsible for procurement of computing equipment, or at least for setting procurement policy, it needs to understand how industry "tools," such as the SPIRIT procurement specification, can be used to achieve corporate goals faster and at a lower whole-life cost.

Marketing and product development

The next major group that can significantly influence success or failure in the service management arena consists of the marketing and product development groups. In provider companies still in the monopoly phase, these groups may have little influence, but in a competitive market, they often rightly drive major decisions involving investment levels and priorities. These groups need to understand the three-legged stool problem more than most, since an imbalance directly affects service pricing and attractiveness to the customer.

Too often, unfortunately, the drive to get new services launched can outweigh the initial cost and service quality imperatives. This issue is made complex by the fact that there might be a significant delay between the day a service is hurriedly launched with handcrafted management systems and the day when the true operating costs and inadequacies of customer service show up. There are two solutions: Get promoted quickly and move on or balance the short- and long-term characteristics of a service launch. If absolutely necessary, a middle alternative is to launch without all of the integration steps in place, provided that there is a plan for coping with the service management issues when the service volumes start to roll in and provided that those costs have been factored into the business case.

Reengineering groups

The final constituency, and certainly not one present in every service provider, is the reengineering group. Becoming more and more common, these groups work best when composed of a multiskilled team drawn from all of the constituencies above. Reengineering teams can be more or less successful depending on the degree of authority they are given by senior management and the extent to which they achieve buy-in to their often radical ideas by those other constituencies. If your company has developed a strong reengineering team, it's a good idea to converse with the members in the early days of your thinking process. You might find that your planned improvement is eclipsed as whole processes are rearranged, or you might be able to ride on its wings to get even better support for your plans.

Dealing with multiple internal constituencies

There are, of course, many other organizations within a service provider. The constituencies described are very generalized and vary considerably from operator to operator. The main point of all of this is that service management excellence cannot be achieved unless all of these constituencies are capable of focusing their energies towards common goals. Too often, energies are dissipated by different groups pulling in different directions.

If you've ever faced the job of pulling all of these different views together and getting an organization to learn to solve problems collaboratively, you'll agree that this is by far the hardest part of your task. By comparison, the technical solutions and pan-industry agreements formed outside your company are much easier to achieve than a unified implementation plan across multiple, competing priorities, capabilities, perspectives, and personalities. But it can be done, and when it is, you achieve something very special—the power of one. In other words, you unlock the energies of large numbers of people behind one single approach rather than dissipating them in multiple, different directions, as shown in Figure 20.1.

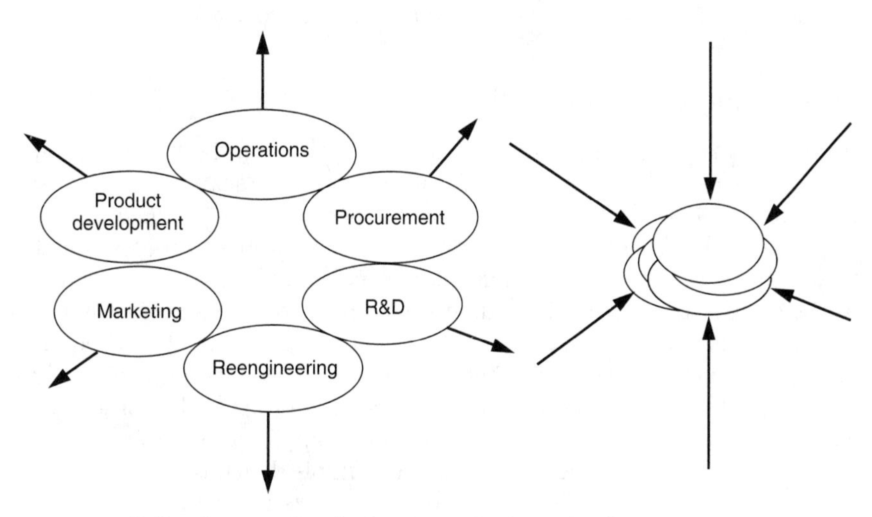

Figure 20.1 Pulling the same direction keeps energies focused and strong.

Beware the stovepipe

If there is one result of a lack of shared vision and strategy that will cause a service provider to fail to survive industry shakeout, it is where different operational support and management systems are deployed, requiring different operational processes for each service. Although initially fast to market, the number of systems mushrooms as new services are piled one on top of the other. Time to market may be impressive, but operating costs will not be. Unless the service development plan includes investment funding and a strategy to bring the overlay into a common architecture, the operator will be saddled with long-run inefficiencies and loss of competitive advantage against smarter rivals.

This overlay approach is sometimes called the "stovepipe syndrome," because each and every service has its own unique set of systems and processes, like so many stovepipes belching smoke into the air. Communications is primarily a commodity business, where economies of scale drive down the cost base. Breaking operational processes and systems into these stovepipes nullifies any economy of scale and puts a major provider at significant risk of losing the advantage to niche players. It may be impossible to avoid this happening occasionally—sometimes the time to market really does override all other considerations. But the three-legged stool calls for low operating costs, high levels of customer service, and short time to market to be delivered simultaneously if service management excellence is to be achieved. Only when every constituency agrees to build an infrastructure that is capable of tight integration will that be possible. Without a common company plan—product development, process development, procurement, R&D, and computing policies all lined up—the operator will fail to achieve the level of excellence that another, better-organized service provider will deliver.

External Constituencies

Whom you deal with externally depends on what you are trying to accomplish. An improvement that ties service management processes together across company boundaries involves working with other providers, or customers, or both. An improvement that connects service management processes with the network management and element management processes necessitates working with communications equipment suppliers or with software houses that may produce systems that help to integrate specific suppliers' equipment. Regardless of the type of improvement you make, you probably also need to work with your computing suppliers, either to provide a base management platform upon which to develop a new application or to meet systems integration requirements to interlink multiple domains.

Although your approach to each of these constituencies is different, there is one consideration that should be made in every case: If possible, don't limit your discussions to a single company. The communications industry is, by its nature, connected to lots of people. Your network infrastructure is connected ubiquitously, and since you have many customers and many suppliers, your service management infrastructure also needs to be connected ubiquitously. Think about others that might benefit equally from such an agreement, and extend your negotiations to include them. It is to your benefit to avoid agreements that are individually tailored.

Working with other service providers

Relationships with other providers may follow a fairly strict customer-supplier model or may be peer-to-peer, in which the customer-supplier roles change with each transaction. Examples of customer-supplier relationships between service providers include:

- The current predominant relationship between the U.S. local and interexchange carriers, where the interexchange carriers are virtually always customers of the local operators.
- Value-added service providers that do not operate a physical network and whose relationships with other providers are therefore always as a customer.

Examples of peer-to-peer relationships are:

- Formal alliances, in which partners support each other as needed to complete an end-to-end service.
- Occasional-use relationships, in which providers may buy capacity or geographical coverage from each other at tariffed rates, just like any other customer.

Customer-supplier considerations. Many service providers take pains to avoid the immediate reaction of their employees to consider another provider "the enemy"—particularly if that other company pays a lot of money for services each year. Where volumes of interactions are very high between provider companies, special groups are often established to deal with other providers as wholesale customers, and the processes and systems used to manage those transactions are separate from the rest of the company's processes, making it easier to negotiate a move to automated process flow-through.

Where volumes are sporadic, the chances are that mainstream systems are used to handle orders from another provider, just like any other customer. In this case, making a change could be more difficult, since your requests for an automated interface may be considered in the same way that such a request would be considered if received from a government account or any other major enterprise.

In either circumstance (but particularly the latter), your case for automated process flow-through can be strengthened considerably if you propose an agreement that has the support of a wider group of companies (that's where the OMNI*Point* Solution Sets can help). No matter how eager another provider is to court your business, it has other provider-customers as well and won't want to support unique agreements.

Equally, it is to your benefit to negotiate with multiple provider-suppliers at the same time, assuming that your company relies on more than one subcontractual provider relationship to do business. And one further caution: A relationship that starts out being one-way may turn into a two-way exchange when the regulatory environment changes. One U.S. local exchange company reported that an interexchange carrier was adamant about getting automated access to information when in the customer role, but as competitive constraints began to be lifted, it was unwilling to provide that same level of information back to the local exchange company!

Peer-to-peer negotiations with other providers. Alliances represent an interesting challenge. On the one hand, an objective of the alliance is to do things better than any other alliance, so there is a natural tendency not to combine forces with others outside the group. On the other hand, it is hard to predict how long an alliance might last or whether it will grow to encompass other companies. Highly tailored agreements (built around existing proprietary service management systems) can be very hard to duplicate, or they may unravel if circumstances change.

Other peer-to-peer negotiations—those that intend to automate exchanges between companies whose roles change dynamically—may be fairly simple to negotiate if approached from a perspective of mutual benefit and equal dependency from the beginning. Competitive issues aside, once two companies agree to do business electronically, both should find it advantageous to implement standard interfaces. Doing anything else risks exposing internal (proprietary) processes and increases the cost of implementation. Of course, this cost issue goes for any interface—standard interfaces are far cheaper if you plan to do more than one—but the "low-cost" argument might sell better with a potential competitor than with a partner.

In any negotiation with another service provider, you are at a distinct advantage in that you will be familiar with the types of internal constituencies within those companies that need to be involved in the implementation planning, since your company and theirs are in the same business. You also have a common point of reference and language in the NMF's Service Management Business Process Model, which can be used by both parties to avoid competitive concerns, particularly during the early negotiation phase before personal relationships are established.

Working with end customers and content providers

When decisions are made to offer online management interfaces to customers, product managers may be more interested in establishing a lock-in strategy than in implementing common interfaces that can be replicated across multiple services. This can be superficially attractive, but there are several strong reasons that balance against this desire.

- First, customers aren't stupid. They will spot a lock-in strategy a mile off and will resist a proprietary approach—after all, they may deal with multiple suppliers.
- Second, standardized interfaces are now becoming well-established, meaning that the customers' own systems are likely to be preequipped with a standard-interface capability. Salespeople will have a tough time selling a proprietary solution unless it's either very cheap or incredibly useful.
- Third, lock-in comes from having a better product—one that delivers more value to the customer in terms of price, performance, and functionality than the competition—not from proprietary "plugs and sockets."

If product management takes a more enlightened view, and developing the customer-handling interface is seen as a key corporate strategy by the service provider, then common electronic interfaces (across services and with as many customers as possible) will be the goal. Within the customer's organization, it is usually fairly easy to find one group responsible for managing the delivery of information and network services to end users. Typically, it is within the IT department.

Gaining agreement to jointly implement a common, industry-accepted interface (such as defined by an OMNI*Point* Solution Set) with a company that is struggling to reduce costs and improve internal service delivery will be much easier than selling a tailored agreement. Remember that the service managers of large corporations are interested in almost anything that helps them cope with the diversity of computing and communications equipment vendors they must deal with.

Working with suppliers

The companies that supply the service providers' computing and telecommunications equipment have as many internal constituencies as do the service providers. Once again, these are large companies with multiple departments often disconnected one from the other or playing out slightly different agendas. Three key groups that must be aligned with the thinking of their customers, the service providers, are the sales and support organizations, the product management organization, and the development organization.

Unfortunately for many suppliers, because the service providers constituencies are not fully aligned, each of these three key groups within a supplier have often received wildly differing messages. Investment in new, advanced integrated management technology is expensive, creating a very dangerous situation leading to nervousness over committing major R&D funds to what appears to be a very risky development. Product functionality has been lower and product introductions have taken longer because of this problem. We hope that this book helps to forward common views being fostered between customers and suppliers to unlock investment and speed product development.

Telecommunications equipment suppliers. Service providers do not often buy stand-alone network or element management systems unless such capability is an add-on to existing equipment. They are more likely to specify management capabilities as part of an equipment purchase. These management requirements should take into account not only the needs of the network operations department but also the needs of service managers, who require certain capabilities at the network level to deliver service features.

Negotiating common interfaces between a provider's own management systems and the network or element management systems provided by equipment suppliers should be easy, but often it's not. First, these supplier companies are usually large themselves, and finding the right people to convince is not always easy. Second, implementing common interfaces at this level is complex and almost always involves considerable investment (and therefore product support and approval within the supplier company) to implement.

Negotiations can be helped tremendously if the request for common management interfaces is tied to a large procurement of equipment, although there are drawbacks to that strategy, too. While you will certainly get their attention, experience has shown that if the procurement is large, lots of people from the service provider's organization become involved in the negotiations. In the final analysis, other considerations (price and features) may take center stage in the negotiation process, and the request for a common management interface may be lost in the noise.

You can't afford to lose this point with your suppliers if you're really going to become a lean service provider. The costs of failure won't show up initially, but they will be there, week in and week out, as higher operational costs if equipment you purchase does not integrate tightly into your overall, highly automated service and network management infrastructure.

If you have to concede ground, go for a phased implementation of a standards-based approach rather than a proprietary offering with the promise of standards later. The latter is fatal—the standard version never comes, or if it does, you will have built up so many software and human work-arounds to fit the proprietary approach that you may be reluctant to touch it. Better not to start—go for a multiphase approach where you get, say, event and alarm information on day 1, configuration capabilities in phase 2, performance information in phase 3, and so on.

As we said earlier, a favorite line of some suppliers is that the standards in the management area are unstable and not yet complete. This is nonsense, at least from a practical perspective. Although standards groups will be working for decades to round out every standard that might ultimately be needed, the fact is that there are enough highly specific and complete standards in this area to float a battleship.

However, until and unless the service providers get their procurement act together, suppliers will continue to balk. We are hopeful that with the introduction of the OMNI*Point* Solution Sets, some of the smoke will clear on this issue and suppliers will be forced to either commit or admit their unwillingness to deliver integrated, industry-common interfaces.

The telecom service provider industry is huge, and equipment suppliers are sensitive to trends in which multiple providers request the same thing. It is far easier to prove in an investment that has value for multiple customers. The message here: Service providers can benefit by procuring against the same standards, but only if they are crystal clear about what is needed, as is the case when specifying compliance to OMNI*Point* Solution Sets and Component Sets.

Also, suppliers are unwilling to let a competitor beat them to a major competitive advantage. We have seen many instances in which smaller suppliers see the competitive advantage of delivering standards-based solutions and act on those opportunities to gain a strong foothold with new products. Once they do, there is considerably more incentive for the "big guys" to follow suit.

Computing suppliers. Within the past few years, most computing suppliers have begun to focus on the needs of service providers as a unique market sector. This is good news in the sense that there is usually a single product group that service providers can work with on issues of management platforms and related applications software. However, these sector-specific groups tend to focus only on network and element management systems, and they may not be well-connected with mainstream product groups in their companies from which service providers need to buy general-purpose, well-integrated distributed computing platforms for service management applications such as ordering and billing systems.

This is not a surprise, really, since even within the service provider environment, the responsibility for buying computing equipment and software is often split between network operations and the IT department. However, these two views need to be brought together if systems-level integration and process-level integration issues are both to be addressed as part of a service improvement.

The different procurements that need to be managed by the service provider can be summarized as follows:

- General-purpose, distributed computing platforms provide basic capabilities such as programming language support, user interface, transaction processing, and the ability to communicate with other systems in a common way. Service providers will probably purchase these platforms as servers or clients, some with transactional capabilities and some without. As part of the procurement of a general-purpose platform, service providers also want to make certain that it can be managed and that it supports portability standards. These general-purpose platforms (the subject of the NMF's SPIRIT specifications) are not specific as to the type of applications they can support, and they can be used, in various sizes and configurations, for network management systems, background billing systems, or payroll systems.
- Management platforms are basically sets of software that sit on top of a general-purpose platform and provide support for specific management functions, such as event reporting or configuration. These management-specific platforms are the subject of the OMNI*Point* Component Sets (e.g., the TMN Basic Management Platform) and serve as a common base upon which to build management applications.

Educating suppliers

Service providers aren't usually much help in educating their suppliers—including their equipment suppliers and computing suppliers. Time and time again, we have spoken with suppliers that are totally frustrated with the potential to serve the communications services business and their inability to do so effectively. Repeatedly, we have listened as suppliers have enumerated the different messages they hear from service providers. Service providers don't seem to think it is necessary (or perhaps good policy) to teach suppliers about their business so that the suppliers can better understand what is needed and by whom. Some reasons for this problem:

- Individuals within a service provider company don't know enough about the whole company to adequately explain the business.
- Service providers are nervous about sharing proprietary information with a supplier, for fear that it might end up being passed to other service providers.
- Service providers believe that they are technically superior to their suppliers, developing detailed requirements (including not just the "what" but the "how") and delivering these requirements without any accompanying business rationale or objectives.

The enthusiasm shown by suppliers over the NMF's Service Management Business Process Model has made it obvious that suppliers are hungry for a glimpse into this very important industry segment. Service providers can't achieve service management excellence without supplier partners. When a specific improvement is contemplated, and you need help from your suppliers to achieve it, spend some time reviewing what you are trying to accomplish and what you hope to gain as a result.

Don't waste time trying to tell them how to accomplish modifications to their systems or how to build their platforms. Instead, focus on what it is you need to be able to do or what information you need to be able to see and what it must look like. Be very precise about the results that must be achieved from process flow-through and what your service management systems expect to receive from theirs. If a Solution Set or Component Set fits the need, require compliance to its terms, and let the suppliers figure out the best way to make that happen.

Chapter

21

Choosing a Practical Implementation Approach

Depending on which problem or problems you decide to address first, you are facing a procurement effort, a development effort, a systems integration effort, or a combination of all three. For example, are you attempting to improve the internal flow of information between your ordering and billing systems as you implement new billing functionality? If so, you are involving internal applications development and also need to think about your systems integration capabilities. Your company's IT procurement policies may well be affected as you develop the new billing capability in a way that works across existing and new systems.

Are you looking to automate trouble ticket exchange across your company or with other service providers? Then you and any working partners need to make changes to existing systems or undertake new development, in which case the computing platform is also a consideration.

Do you wish to offer a new service, such as an on-demand bandwidth service, which can be reconfigured using common commands regardless of the type of networking technology used to provide the broadband capability? If so, you need to develop the appropriate service management applications in your systems, while also driving changes to the equipment suppliers' network management systems through procurement actions.

It is important to keep in mind that the types of agreements reached by industry address only the "plugs and sockets" where the systems of two entities meet. Actually making use of industry-common information that can become available through such an interface—by linking the bandwidth configuration function together with real-time capacity checks, authorization checks, and billing instructions, for example—can be much harder. And because this internal linking of systems goes to the heart of service competitiveness, there is little the industry at large can do together to solve a specific company's problems, other than providing all of the architectural building blocks that help foster systems-level integration and multipurpose management platforms to make the job physically easier to implement. So, although every situation is likely to be unique, and there are few off-the-shelf solutions to help, a number of general frameworks, approaches, and standards are relevant.

In this chapter, we look at some of the options open to systems developers when planning to implement process flow-through interfaces between management systems using common industry techniques.

There are three ways in which common management interfaces can be applied: federation of existing systems, integration of applications on a common distributed platform, and a hybrid that uses a combination of the two. Which approach you choose to employ depends on your own environment and can vary across service providers and suppliers.

Dealing with the Legacy Environment: Federation as a Logical First Step

As cable TV or mobile communications companies can attest, service management is not an easy thing to do even when a company has a "green field" in which to construct an all-new set of systems. Consider then, the added complexity of achieving flow-through across systems that have existed for decades independently of one another.

Traditional service providers (which also operate networks) have hundreds of legacy systems to support their many customer care, service development and operations, and network management systems. Typically, the element management systems are provided by third-party suppliers, while the network and service management systems are often home-grown. Both present problems when trying to interoperate and produce automated process flow-through.

The federation of systems—bringing together diverse systems and making them seem as if they were one, at least for certain functions—is particularly appropriate when trying to implement a common interface across multiple proprietary systems. In the TMN lexicon, this is known as using mediation devices. In essence, the federation approach makes each legacy system look like a "black box" to which a common interface "gateway" has been added. The gateway software communicates with other systems in accordance with the interface agreement and then "translates" the common information into whatever language and terms are understood by the legacy system. Service providers in the United States that have implemented the OMNI*Point* trouble administration interface have employed a federated approach as a way to automate a critical function without modifying their underlying trouble administration systems or processes.

Figure 21.1 shows how the federation concept works and how it might be applied by different companies to implement automated interfaces at all levels of management—element, network, and service.

Federation does not provide a totally integrated solution, and many compromises and choices have to be made. It is helpful to use the Pareto Principle in driving these choices. In any service provider process, 20% of system interactions are likely to account for 80% of the cost. By focusing on federating the most regularly exchanged information, major inroads into cost reduction and process flow-through can be made by this method.

Distributed Platforms: Taking Advantage of Client/Server Technology

Some service providers, notably the "new breed" of value-added providers—the cable industry and the cellular providers—have the advantage of a clean slate when it comes to their management systems. They can build distributed management systems using platforms that provide a facility for common system-to-system communications, as well as key services that can be

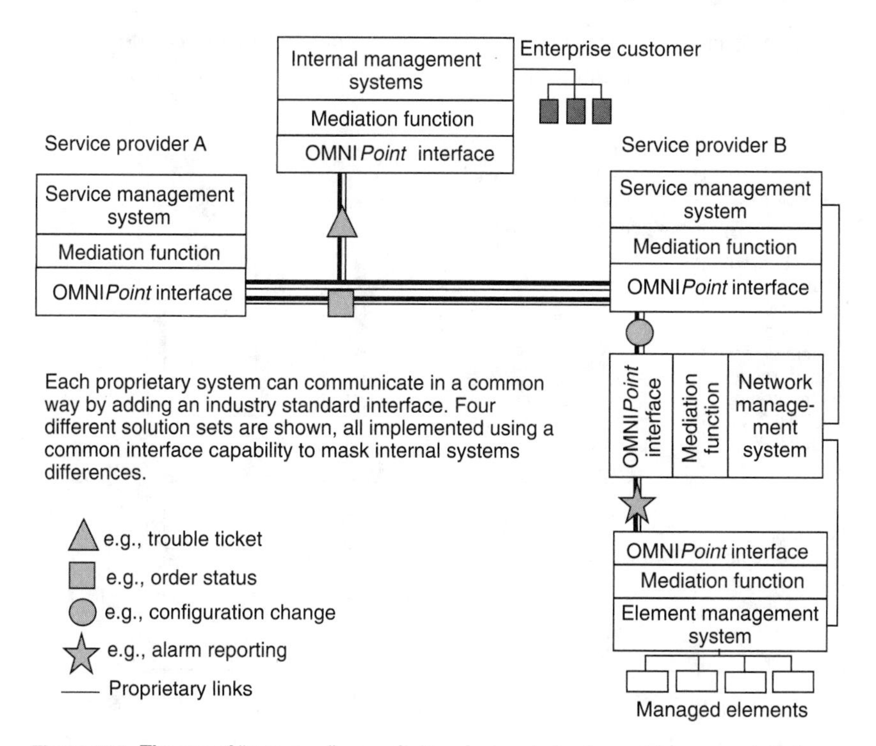

Figure 21.1 The use of "gateways" or mediation devices to implement a common interface.

used by the applications, thus reducing the amount of new development each applications developer needs to do.

Figure 21.2 shows how this distributed approach might be used in developing a range of management applications to support service, network, and element management needs. It also shows how a company might use the SPIRIT specifications as a base and then apply appropriate management applications support capabilities through compliance with the OMNI*Point* TMN Basic Management Platform.

This same concept—use of a common software platform—has been used extensively in enterprise networks to implement the Internet management model, whereby common network management platforms provide SNMP communications capability and a MIB (Management Information Base) access facility, and applications developers build specific solutions that use these facilities.

The idea here is that most communications between applications is conducted "inside" the distributed (client/server) computing platform, using common computing techniques such as DCE to exchange information directly, without the need for the type of TMN interface that is frequently used in network and element management applications. There are several advantages to a distributed approach:

- Applications are separated from the underlying computing technology by applications programming interfaces (APIs), making development easier and faster.

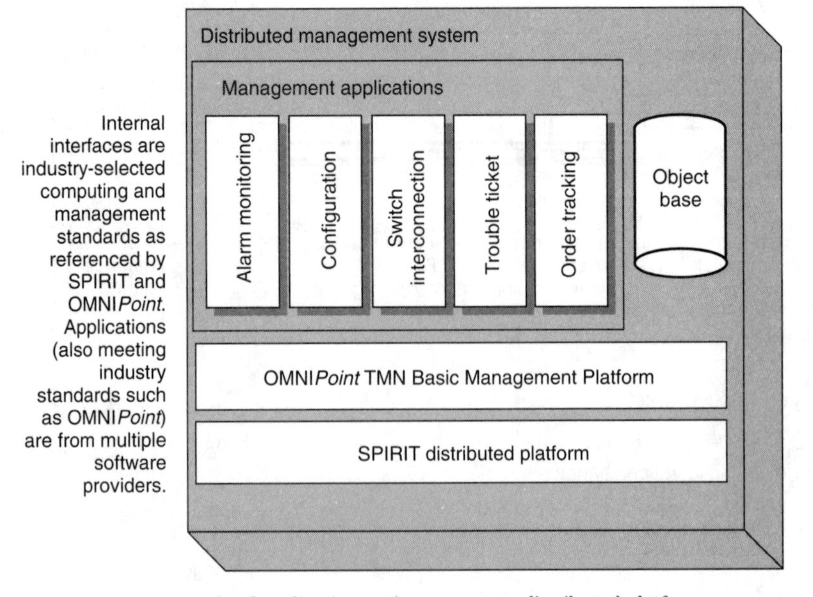

Figure 21.2 An example of applications using a common distributed platform.

- Applications development itself can be distributed and undertaken in parallel, unlike traditional, monolithic systems development, which is serial. This considerably reduces the bottlenecks or queues that often accompany new service introduction and shortens time to market.
- A marketplace for standardized applications packages can develop (and, in fact, is already developing) as APIs become standardized and more useful. Rather than develop all of the applications uniquely, service providers that use standard software modules can cut development time scales and expense. This has other benefits, as well, including reduced user training requirements. Rather than spending time learning how one systems environment differs from another, users are free to concentrate all of their attention on the application, and the use of common graphical user interfaces, common icons, and common command structures means that users can become proficient with an application much faster.
- A distributed approach gives a high degree of built-in data and application integration, providing the maximum opportunity to achieve high levels of process flow-through.

The concept of using a common, industry-supported management platform, completely interoperable with like platforms and permitting customers to choose from among a wide range of compatible management applications, has a great deal of appeal. But the jury is still out regarding its feasibility. In the enterprise market, service managers are becoming frustrated, saying that in all the years of trying to achieve the goal, suppliers have not yet made their own platforms interoperate, much less achieved interoperability with another vendor's product. What's more, the levels of support offered to applications has been extremely weak. Perhaps the concept of a single, industry-supported management platform borne out of the abortive attempts of the Open Software Foundation to create a distributed management environment is like so many press-hyped solutions—a "magic bullet" that doesn't actually exist.

According to Jim Herman of Northeast Consulting Resources, two different models are emerging to bring some reality to the management platform model, at least in the enterprise market. One is that vendors will collaborate with software suppliers to produce "suites" of applications along with a platform. Under this model, buyers might not get "best of breed" applications, but neither would they need to worry about making the different software products work. One assumes that under this model, the underlying platform would be able to interoperate with other installations of itself but not with other vendors' products.

The second model is that of solutions—in effect, a customized packaging of applications and platforms offered by computing suppliers acting as systems integrators. Here, the customer can pick and choose from among applications but would pay a vendor to integrate the various offerings. In essence, the solutions model simply shifts the problem of management systems integration back to the computing supplier—much as the problem for communications service management has been shifted back to the service providers.

These approaches, which might coexist, seem a possible approach for the telecom market as well, based around a common set of computing platform choices such as that described in the NMF's SPIRIT documentation. Here are a few reasons:

- Service providers, which have considerable IT expertise in-house, have already implemented management platforms based on standards that they have developed on their own to support multiple service management applications. They know it can be done and they expect their suppliers to replicate (and improve upon) their experiences.
- Service providers have, in the past, relied less on external management applications development than have enterprise customers. So their critical need is getting a common communications ability and a set of common management services (event reporting, state management, etc.) that can serve as utilities to support internal applications development. This reduces the dependency on one of the critical "plug-and-play" elements considered part of the open management platform concept.
- Service providers want to move to a distributed, object-oriented environment for new systems development. Their networks are simply too large and their personnel increasingly too spread out to manage without such tools. Service providers have already experimented with early versions of such technology and are working closely with computing suppliers to shape future products.
- Service providers, more than enterprise customers, are able to focus their attention on the few key standards they must have in common from all computing suppliers. By giving the vendors a clear target to shoot at, it is possible that interoperability among distributed management platforms can be achieved sufficiently for service providers to get the type of systems-level integration needed to support process-level integration.

Assuming that service providers are successful in achieving some level of platform interoperability across computing suppliers, it should still be remembered that most systems are embedded; that is, they already exist and will not be rewritten, at least not in the foreseeable future. So, the reality of implementation is that service providers will implement a hybrid approach.

The Hybrid Approach: Covering All the Bases

Most service managers inevitably have to consider a hybrid approach, whereby new management applications are implemented on a mixture of common management platforms and legacy systems, as shown in Figure 21.3.

A number of approaches can be taken using gateways that allow the legacy systems to interoperate with standardized distributed platforms. This client/server/server architecture, introduced in Part 3, plays an important role where the legacy system (often built on a mainframe computer) acts as a server to a distributed client/server system.

Although certainly not pain-free from an operational perspective, the client/server/server architecture has major advantages in that the migration from existing databases and applications can be made in a phased manner. New applications that use data resident in the mainframe server can be created outside the mainframe environment, reducing development time (and time to market) by bypassing the development queues often associated with monolithic systems.

Interoperability between computing domains—the mainframe or "corporate" systems, the distributed client/server systems, and the desktop—is another area addressed by the SPIRIT specifications. The OMNI*Point* platform specifications, as they exist now, are really most appropriate for use in linking with network management systems or for network management systems to link with element management systems. Although the same techniques are being used to exchange non-real-time information, as well (e.g., trouble tickets between service providers), we expect that the industry will begin to adopt more generic distributed computing techniques as more and more service management system linkages are required. In addition to the basic platform specification work done as part of SPIRIT, the work of the Telecommunications Integrated Networking Architecture Consortium (TINA-C) may help to apply DCE and object-oriented con-

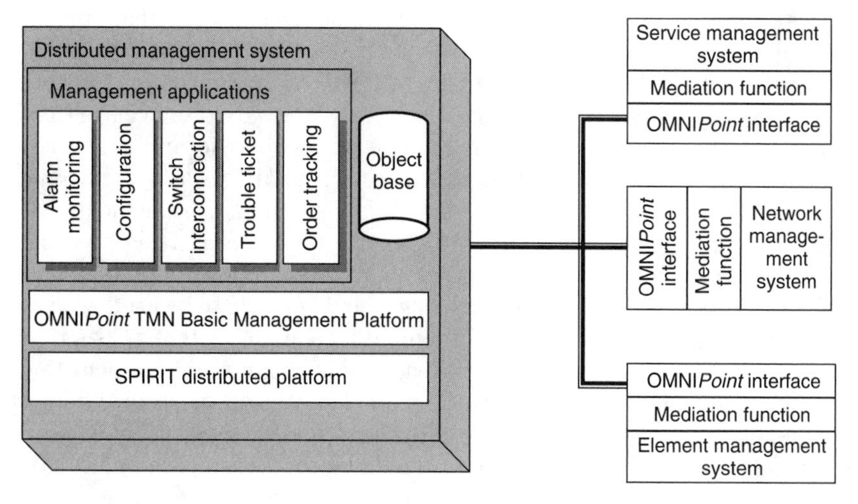

Figure 21.3 The hybrid approach—a combination of legacy and new systems.

cepts to the development of platforms that more adequately fit a service management environment.

Why is this important? Advances in computing technology are key to unlocking the power needed by service providers to manage their enormous networks more easily and at less cost. Remember the term *disruptive technology* that we used earlier in the book? Distributed computing, distributed databases, and object-oriented technologies are about as disruptive as you can get when assessing the impact of new technologies on traditional ways of doing business. Lee Jobe, formerly director of business operations at Sprint, said "By moving to a distributed-operations platform, with the application functionality in the hands of operations people, we cut the introduction of new operational features from months or years to days and weeks. This made a major impact on Sprint's ability to improve its service quality and simultaneously cut its costs."

The model that Jobe presided over at Sprint was somewhat unique in that virtually all applications development for his installation and maintenance centers was done by a premises-based development team made up of IT and operations technicians who were trained to develop with newer tools, such as Microsoft's Visual Basic and Visual C, that are easily learned. Jobe, who is now with Pacific Bell as its vice-president of business operations, intends to further exploit this model because he believes it is the only way to build flexibility into PacBell's service operations centers.

This observation is a crucial one. No longer do operations people or product managers, frustrated with huge bottleneck delays for even the simplest change to a process, need to wait. They can do it for themselves without endangering the system as a whole.

Most service providers acknowledge the need for a longer-term move toward object-oriented approaches to handle their thousands of network nodes, hundreds of services, millions of customers, and events that number several thousand per second— and to keep track of the relationships between these many things as changes are made dynamically. The type of technology needed by providers to streamline the systems used to manage their core networks is only now beginning to emerge and only for use in smaller-scale systems. In the grand scheme of things, the industry has not yet come far enough, and the work done to identify implementation agreements has been bottom-up. But the work that has been done serves as a good starting point for achieving intercompany process-level integration, and the models used to capture the information exchange should be able to be used even as advances are made in the underlying computing technology that will support management applications. Looking ahead, industry groups are anticipating what changes might occur and what technologies might exist. In the next chapter, we take a look at how some of the coming technology might be used to provide the kind of flexibility needed to link processes end-to-end.

Chapter

22

Putting Principles into Practice

In this book we have presented a number of facets of architecture—principles and structures from which to build a coordinated approach to service management. Architectures are all very well for architects, you might say, but what use are they to me? "I'm a product manager" or "operations manager," or perhaps "I'm in sales" or "I'm in marketing." The real value of an architecture is that it provides a common framework and language so that people can communicate ideas and make progress in solving problems. In the words of Michael Scott-Morton, the J.W. Forrester Professor of Management at Massachusetts Institute of Technology, ideas and frameworks "provide a lens through which to view the world and understand complex problems more clearly."

The layered architecture that forms the core of TMN thinking came about simply because people had no common way of breaking down the issues of managing an entire service provider's business into pieces that could be used to focus energies. Concepts such as partitioning, abstraction, and recursion might sound daunting, but they are really just examples of that "lens" in action—helping people describe the real world to each other through a common set of concepts and terms.

In this chapter, we lay out some real-world examples to bring those architectural concepts to life. That's important, particularly if you are not a systems architect. If you're in sales, it gives you some principles with which to understand your customers' requirements; if you're in marketing or product management, a set of tools that enable you to describe your needs to your systems people or suppliers. And if you are a systems architect, you'll know that what we've presented in this book is only at a high level and that you need much more detail to really implement systems.

Architectures and frameworks are at their best when they are simple. They are not rocket science—anyone could develop an architecture given some time. Developing a better architecture misses the point. Esperanto is probably a better language than English—certainly much less idiosyncratic. But because so few people speak it, its value as a communications vehicle is limited. So too with architectures. A proprietary one may be "better" than a common framework but of little use to explain concepts to others. We introduced several frameworks in the second part of this book. Let's look at a practical example to see how they can be used as that "lens" on a real-world problem.

A Real-World Example: "Supercom"

Let us assume that the marketing team of an imaginary service provider called Supercom has seen a growing need to introduce on-demand bandwidth services to supplement its portfolio of fixed private line services. Supercom's product marketing people have produced a product specification that requires the following:

- The service will allow customers to contract with Supercom for a variable bandwidth service that allows them to adjust their bandwidth requirements from a minimum of 9.6Kbs to a maximum of 2Mbs on demand.
- The customer should have full online visibility of the service, including network status, online fault reporting, online problem resolution capability, online ordering capability for additional services, and all of the tools necessary to change bandwidth needs.
- The service will offer a two-minute guaranteed maximum time to effect a bandwidth change.
- The service is to be priced so that, on average, no more than a 3% price premium exists over fixed bandwidth services.
- Because of competitive pressure, the service must be ready for launch within four months.
- The customer management facility must be capable of integration with the customer's own management systems to allow fully integrated control by the customer. Note that the customer may have a wide variety of customer premises systems for use with the service, including local area networks, PBXs, and video conferencing systems.

In that simple service specification, all of the elements of the three-legged stool are in evidence. What is required is a very cost-effective, high-quality service with significant service management features, all to be delivered in a short product development interval. Supercom's ability to meet that service specification depends entirely on where it is in the evolution cycle of becoming a lean service operator provider.

Two significant things box in the designers:

1. If the service is implemented using people to deliver the service functions, the network operating cost will probably exceed the service price—not an attractive proposition! Using manual methods, the two-minute service level guarantee could probably not be attained.
2. On the other hand, this is a highly complex service, and automation of the functions demands considerable interaction between a number of element management, network management and service management systems. Unless Supercom has begun to invest in some standardization of those systems, service development time scales may very well be four years rather than four months!

Let's pick this service requirement apart and use it to try to illuminate some of the architectural principles we reviewed earlier. What we have here is a service that is fairly straightforward at the network level. Supercom has a modern network using synchronous digital hierarchy (SDH) as a backbone, and, in addition, it uses modern time division multiplexers (TDM) on customers' sites to deliver service. Both the SDH backbone and the TDM access network have fairly good element management capabilities, which allow bandwidth to be assigned dynamically from remote management equipment, although initial installation requires a visit to the customer's site to install appropriate line cards, etc. Supercom also has a number of other network-level management systems, including systems that track and assign network capacity, systems that give overall network status by combining the alarm outputs of several element management systems, and a system it is working on (but has not yet deployed) that will integrate the configuration management tasks of a number of network resources.

At the service layer, Supercom has a variety of management systems. Billing is implemented on several systems, one of which, a 1980s vintage mainframe system, is used for private line billing. Various other systems exist, and these are separate for functions such as customer administration, fault histories, and ordering. Supercom is in the process of rationalizing its service management layer, but, at present, these ideas have gone no further than a series of concept papers passed to numerous internal committees for consideration. To date, Supercom has no service offerings that provide full customer control or visibility.

Supercom could be any one of a hundred different service providers with pretty good network technology but surrounded by a hodgepodge of systems and processes deployed around it. It's passable for business as usual, but it falls short when the company introduces a new service like the one outlined. Let's take a look at some of the challenges that our design team will face and the bad news that they may have to take back to their marketing colleagues.

Network to element level integration

Looked at "bottom up," the situation isn't too bad. At least Supercom has a network capable of delivering the basic function of the service. Bandwidth can be assigned dynamically across both the backbone and customer access networks. The first problem occurs here, however, since there is no tight integration between the management of the access layer and the backbone. This means that when configuring that bandwidth across multiple network elements on behalf of the customer, human beings will have to be deployed if the service introduction target is to be met. "OK," say the marketing guys, "we can probably live with that as a phase I product." The downside is that the two-minute service level guarantee might need to be extended to 30 minutes to allow the technicians to run from management console to management console to implement network change requests. For phase II of the service introduction, the development team will have to accelerate its plans for introducing an integrated configuration system between the two element management systems.

Let's take a look at the service proposition so far, using the layered architecture and the management systems framework introduced in Part 2 of this book.

As shown in Figure 22.1, the proposed network configuration management system is key to implementing this service. This system needs to be integrated not only with the element management systems but with other network-level systems, as well. Its job is not to replace the element managers but rather to provide an integrated view of the network configuration at the network level or, as in the case with this variable service, to provide an abstracted view of the network to the service level. The reason why the view needs to be abstracted (i.e., at a higher level and with less detail than is

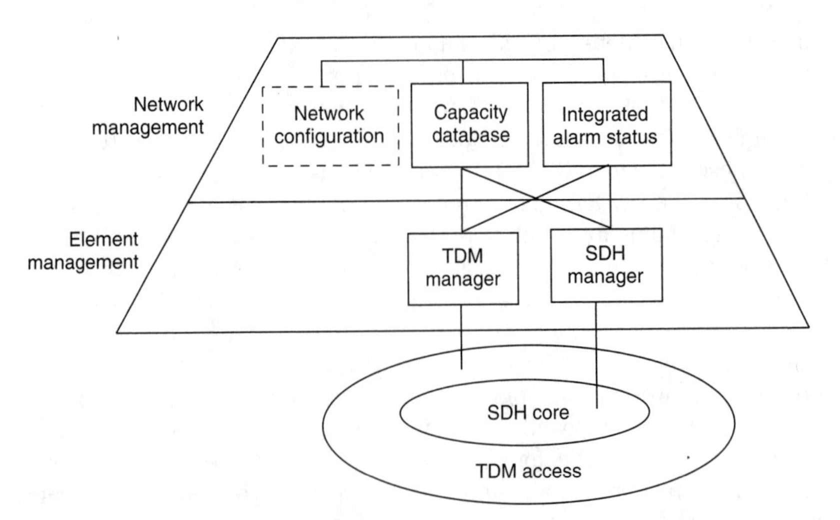

Figure 22.1 Supercom's network and element management systems.

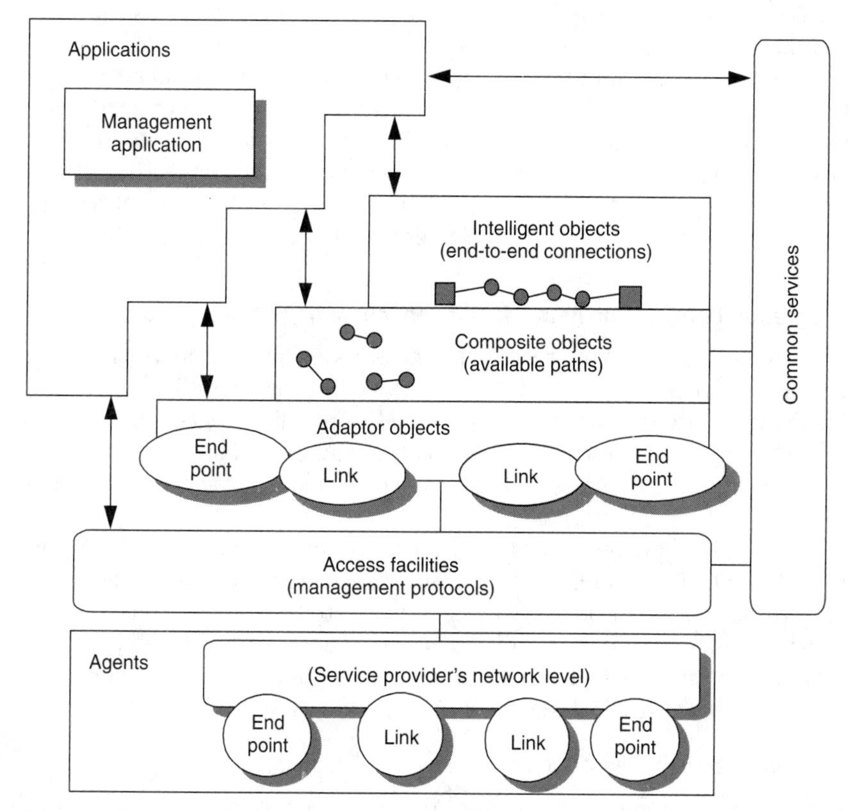

Figure 22.2 Applying the management systems framework to Supercom's problem.

needed at the element level), is because the network needs to be viewed not as a series of subsystems but as a series of links with end points—in other words, pathways across the network rather than the implementation detail of how those pathways are constructed. Seen from a customer's point of view, that's all that matters, since it is what he or she is paying for. How the network is routed physically is of little interest from the customer's standpoint.

To achieve this feat—looking at the network in increasingly abstracted terms—the systems designer needs to make use of object orientation. There are several opportunities to describe the network resources and their surrounding management capability as objects, from the simple adaptor object described in the management systems framework overviewed in Part 2 through intelligent objects that achieve the highest level of abstraction between the network management layer and the element management layer. Figure 22.2 gives some idea of how object encapsulation might be used.

Adaptor objects are used to represent the links and end points. Composite objects provide an abstracted view of these resources in terms of actual paths that are available in the network. Intelligent objects make use of these composite objects to build an end-to-end connection.

From the viewpoint of the configuration management system at the network level, the entire SDH backbone and the TDM access layer need to be modeled as one resource capable of reporting information of use to the next layer and taking instructions to set up specific links between end points on the network. Beyond that, the detailed information about the SDH and TDM components are held at the element level.

Integration between the network and service management layers

Having the capability to reconfigure network elements from the network level is the crucial starting point for a service like this, but it is not enough. Facing the network from the service level, something is needed to direct the network configuration system to make changes that correspond with individual customer requests and that can keep a current record of each customer's service configuration.

From Supercom's perspective, the link between network capabilities and individual customer services is very important. Supercom doesn't just want to have one customer on this service, it wants to have many. Thus it needs some sort of service configuration system to facilitate the dynamic reconfiguration of the service by the customer or by Supercom's customer operations personnel. To be effective, it must provide several things:

- An abstracted view of the entire customer service.
- Security—methods of preventing the customer from gaining access to systems information to which he or she is not authorized.
- Partitioning—allowing the customer access only to information relevant to that customer.

The biggest reason for providing integration between the service and network levels is the need for partitioning. The service specification requires that the customer can control a part of the network at will. Partitioning is essential if customer control is to be implemented, since the scope of customer control must be limited to only that customer's slice of the network. It would be a disaster if customers could implement changes that affected other customers' service levels or do anything that would put the entire network in jeopardy. Thus partitioning is a major factor in providing security.

Integration within the service management layer

So far, we have only described the network-facing integration issues at the network and service management levels. Before this service can be launched, however, the systems and process design teams have to work out how to take an initial order, how to store information about the customer, how to deal with issues that arise in resolving problems with the customers' service, how to monitor that the service level agreement is being met, and, most importantly, how to ensure that the customer's bill accurately reflects the bandwidth that is being used.

This requires considerable integration to achieve process flow-through within the service management layer. It may be acceptable to the product management team to have a reasonable degree of manual setup for an initial instance of the customer's service. For example, signing up a new customer could involve manual entries into the billing database, problem handling system, customer service configuration system, and so on. But relying on these manual entries won't work when the customer starts making dynamic changes to the service, especially if the two-minute service level agreement is to be met. If all of the systems have been set up for static private line services, then the ability of the customer to vary that bandwidth could throw these systems, especially the billing system, into complete confusion. And what if the service is as popular as the marketing team hopes—could those inefficient manual processes cope with demand from thousands of customers?

This presents the systems designer with a considerable problem if Supercom is not very far along the road of integrating at this level. Just think about the information flows that are required for the simplest of reconfigurable services, as shown in Figure 22.3.

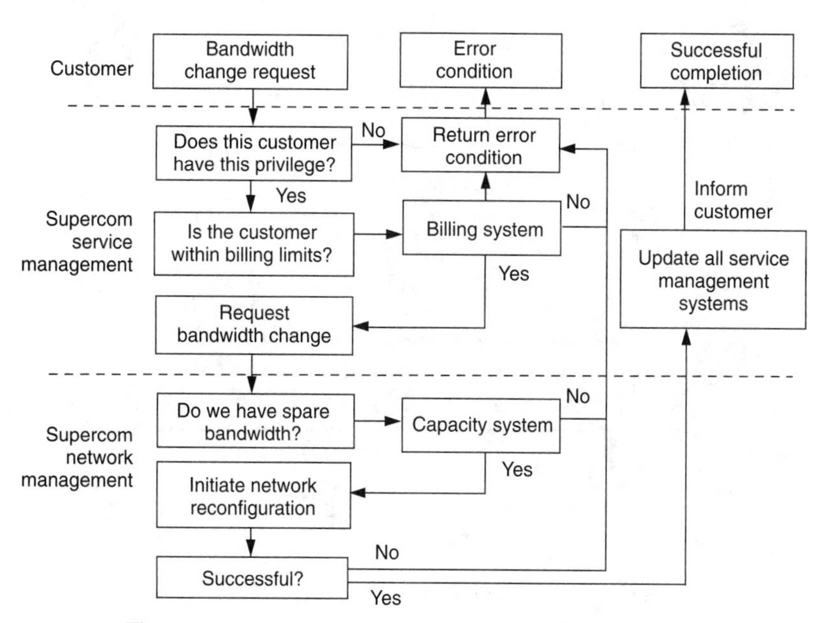

Figure 22.3 The interactions that surround a bandwidth reconfiguration.

Information must flow between service management systems, between the service management layer and the network management layer, and finally onto the element management layer to achieve success. This is an example of the horizontal and vertical integration that we introduced earlier—integration horizontally within a layer and vertically between layers. Let's see what that looks like in terms of the management hierarchy, shown in Figure 22.4.

Supercom-to-customer integration

Everything we've discussed so far is internal to Supercom. Those of you who have actually attempted this task know that we have grossly simplified many of the issues in order to bring out some of the key concepts. We haven't talked, for example, about the physical systems implementation within these layers, such as how to synchronize databases, what happens on failure and restart of one of the systems, user interface design, and on and on. Nevertheless, the internal systems within Supercom are still not the whole story. The outlined service specification calls for full customer control with online status displays, order handling, and problem resolution. The marketing team also wants to ensure that the customer can integrate this information into his or her own management systems to get an inte-

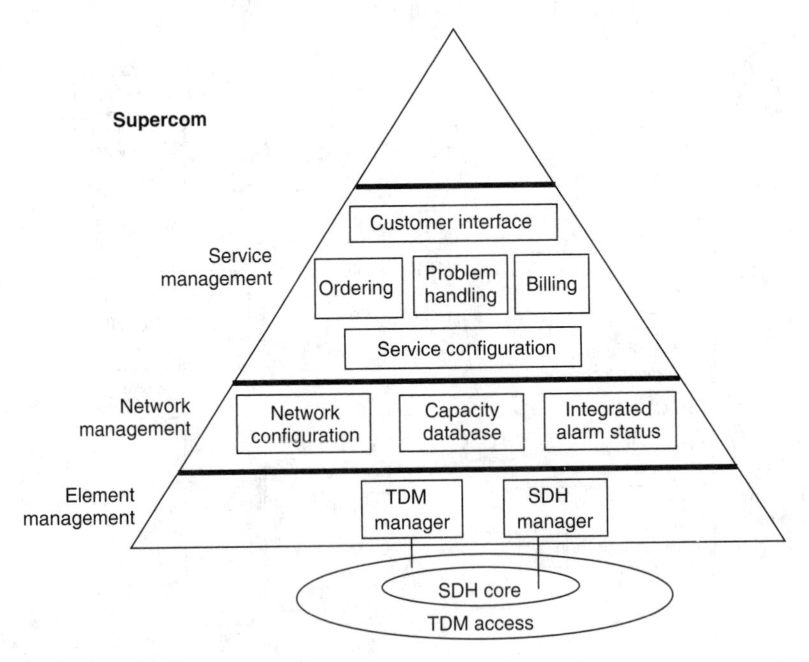

Figure 22.4 A hierarchical view of Supercom's systems interactions.

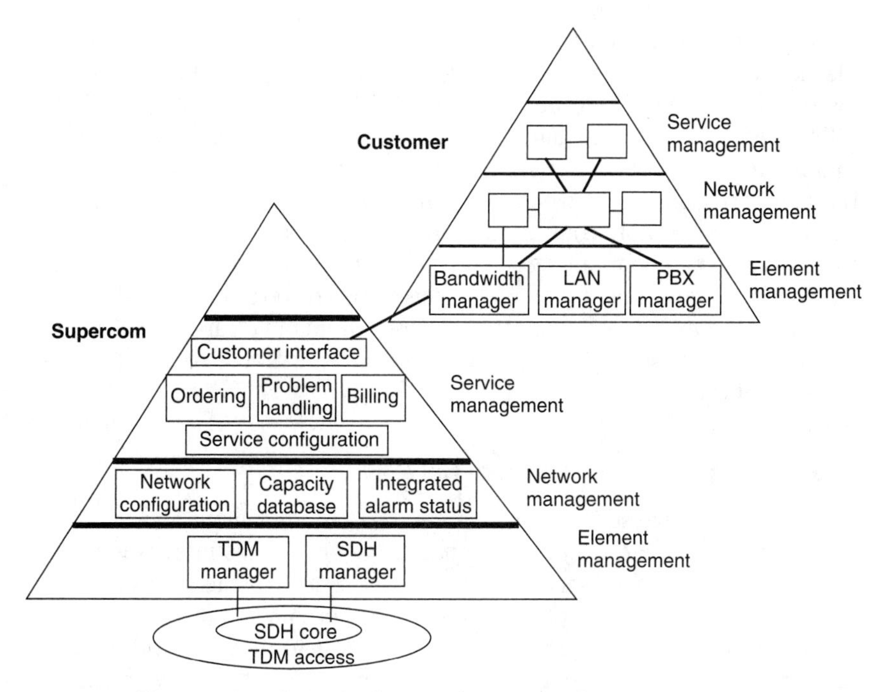

Figure 22.5 The recursive relationship between Supercom and its customer.

grated and composite view of the entire infrastructure as seen from the customer's perspective.

From the customer's viewpoint, all of the sophistication and complexity of Supercom appears as just a humble bandwidth service—admittedly quite a clever bandwidth service in that you can turn it up or down, but in the grand scheme of things, pretty simple stuff. From the customer's perspective this is nowhere near as complex as managing a whole slew of local area systems, complex applications, routers, bridges, PBXs, irate end users, and so on.

At the customer's end, Supercom's variable bandwidth service is just one more element in the universe of technologies that the customer has to manage. It looks just like one more element, and the service management capability offered by Supercom looks like an element management system. In other words, the customer's management picture is a recursion of Supercom's—all of the complexity inside Supercom has been abstracted to the point that it is just an element in the next recursion. This idea is shown pictorially in Figure 22.5.

The customer needs to integrate the management capability of the variable bandwidth service with the output of other management systems. This need almost certainly means that Supercom's service must support a standardized management interface, and any software it supplies needs to run on a standardized management platform.

The most common customer premises management protocol is the simple network management protocol (SNMP). However, within Supercom's own domain, it may well be following a TMN approach using the X.700 (CMIP) protocol series to gain the functionality and sophistication required to integrate its systems. Thus the customer interface system must support the ability to convert between different management protocols. This isn't just a matter of simply converting from one information stream to another as, say, between communications protocols, because management streams contain knowledge and meaning that has to be converted in format as well. It's a bit like working between two different human cultures—not only do you have to translate the language, you have to translate the intended meaning of the words, as well.

Where Does Supercom Go from Here?

From what we have seen in this chapter, even to implement a fairly simple service requires a large amount of management systems integration if the three legs of the stool—service costs, service delivery time scales, and service quality—are to be kept in balance. If Supercom has made the investment in moving down a path towards being a lean provider and reaching high levels of service management excellence, then this service request from its marketing team presents few obstacles. However, if it hasn't made that leap to even understand why such an investment is necessary, then implementing the service will be a lengthy and complex process.

In a monopoly world, that wouldn't have mattered. The service would have taken as long as it took to build and would have been priced at a price that the service provider thought appropriate to recover its investment costs. But as we have stressed throughout this book, life's not like that anymore. If Supercom's competitor does have that capability and can launch this service, it will have a competitive advantage over Supercom that is hard to match. If Supercom were an existing, ex-monopoly provider with 95% market share of the private line market, just think what a new competitor could do to that market share if it could introduce a service like this at or about the same price as a "vanilla" private line service. The cost to Supercom's bottom line of not having implemented an advanced service and network management infrastructure could be measured in hundreds of millions of dollars or deutsche marks.

Given Supercom's situation, the first meeting between the systems designers and product marketing could be a stormy affair. It might not be possible to meet the four-month deadline—the service might take a major amount of time and investment to complete if the foundations have not been laid. If they have, fortunately for Supercom, quite a lot of work has been done within the industry on how to build the necessary objects and systems for this type of service. Managed bandwidth is one of the OMNI-

Point Solution Sets developed by the NMF. Quite a few vendors of both networking technology and computer systems are becoming familiar with this type of problem, and Supercom might find it can buy technology and avoid building everything in its own labs.

So where does Supercom go from here? If it is unable to implement a fairly simple service such as the one we've described, how on earth can it cope with advanced multimedia services? How would it cope with an end-to-end, seamless global version of the managed service we outlined? How would it cope if the regulator broke the company up into pieces such as a networking company and a service company? These are not silly or wild questions; they represent what's happening in the telecommunications industry today. Supercom simply doesn't have the time to spend years engineering its way out of this problem, and the sooner it recognizes that it needs to get a grip on its management infrastructure—both its systems and processes—the sooner it can get itself to a position in which its systems designers can say to product marketing "No problem!"

Chapter 23

Wrapping It Up and Taking It Home

We have flooded you with information and guidance in this book, and we've tried to share with you both the pressing reasons for urgency and the practical steps you need to take to respond to competitive threats. Perhaps a brief summary of where we've been would be helpful at this point.

The Business Imperative

The single most important thing is to recognize what is happening in the communications industry. The changes we are experiencing now are just the beginning, because of a simple fact: This is among the largest, most profitable, and fastest-growing industries there is, and it is being set free from a century of government constraints. Once old protections are removed, a whole new set of players will find ways of profiting from liberalization. Just because a company has experience in managing communications services does not mean its customers will continue to be loyal as others prove they can do it, too, and for a lower price. Remember what AT&T found out about customer loyalty—just being good is not good enough to retain customers. (See Figure 18.1, Kordupleski and Vogel 1988.)

Some companies still have their heads in the sand. These ostriches are sometimes dressed in eagle's feathers, but their words and actions belie their true colors. Others are in the early stages of understanding competitive markets and haven't yet realized that sometimes it pays to work with a competitor in well-defined areas. We've heard a lot of statements that illustrate these points. Here are a few favorites.

- "We don't have to worry about this problem yet." Most companies that make this type of statement have failed to see that competition has already begun to eat away at every part of the communications market. Callback companies, described in Part 1, are a definite part of this trend, breaking down traditional service borders without so much as filing a tariff. And complacent service providers should be more than a little worried if they are following the movements of companies such as Microsoft and Intel. These are tremendously innovative companies that come from "outside" and therefore may see opportunities that are not obvious to those that are too close to the existing infrastructure. The sale of cellular licenses, the entry of the cable industry into telephony, and the intention of entertainment companies to control customer access—these are all indications of a level of competition that many companies simply refuse to recognize and act upon.
- "Service is our unique selling proposition, so everything we do is proprietary and confidential." Well, yes and no. We've said repeatedly that companies will stand or fall on the level of mastery of service management excellence. But doing it all on a proprietary basis is naive in the extreme. Boeing and Airbus produce very different aircraft, but they use many common suppliers and technologies. The costs and time scales of a fully home-grown approach to service management will soon kill off this idea. Even if it could be made to work internally, when it comes to interfacing with customers or other suppliers, they just won't support a proprietary approach because of the costs they will incur and the fear of lock-in.

  Service providers that can deliver information that meets a common, accepted format (so the customer can easily interpret and use the data) using a common, industry-standard interface (so the customer can create automated links with its internal systems, without having to create multiple bilateral interfaces) will be the winners. Differentiation will be based upon the accuracy and quality of the information given, the quality of the services offered, the added features a service provider makes available, and the competence of the employees in handling customer inquiries and complaints.
- "We will never work with the competition." That statement was made by a senior executive of a would-be global service provider when asked whether his company planned to support industry initiatives to establish common links between service providers. It is repeated here because it points out the need to make a rational distinction between those areas in which service providers will (and must) compete and those where it only makes sense to work together. The ability to compete and collaborate si-

multaneously is a mark of maturity that is common in the computer industry but lacking in some parts of the telecom business.

To illustrate the point, Viesturs Vucins, president and CEO of Uniworld, the joint venture of Unisource and AT&T, summed up the dichotomy of today's communications environment in a recent keynote address. He said, "It is our job to fight each other to get customers, but we must also work with each other, because more and more we are each others' customers as well."

Vucins understands the value of industry agreements that make it easier to move information between companies, even if they are fiercely competitive:

> Partnerships are forming and stabilizing, others are breaking up and reforming in different groupings. A clear structure has yet to be established. However, this fermentation will not take too much longer to complete. At the end of it, I foresee a situation in which there remain perhaps three or four main players in the market. Interoperability between these grouping will be commonplace—the [customers] will see to that.

- "We're really focused on our key competitors." This statement was recently made to one of us by a representative of one of the U.S. RBOCs, citing the loss of share to an alternative access provider. Although it is always preferable for a company to recognize where and when competition exists, there are times when too much focus on the major competitor can lead to unexpected consequences. We applaud this company for not ignoring the trends, provided that it doesn't base its forward strategy on this one experience. You will recall the example we cited of the company in the "headless-chicken" phase that focused all its attention on beating competition in the residential markets, meanwhile letting its data services market evaporate. Similarly, an ex-AT&T executive recalls the early days of competition and overreaction, when his home company, Illinois Bell, lodged complaints with the Public Utilities Commission and the FCC about a company called MCI. Illinois Bell took great pains to point out what was wrong with MCI's network design and how routes were being duplicated, no doubt helping MCI to correct its engineering problems, giving it a better start!

  The point is that companies need to exercise caution when developing a strategy for the future. Recognizing who the competition is—both now and in the future—is important, but when the protections are first removed, there is a real temptation to tilt at windmills and waste enormous energy fighting the wrong battles. Don't forget the observations of Kei Takagi that we noted earlier—organizations can get too focused on beating the competition at the expense of meeting their customers' needs. It's easy to get these two priorities muddled up. Of course, watching your competitors is important, but it can't be allowed to subordinate a passion for the customer.

If your company has still not faced reality, then your approach to service management excellence is probably better accomplished by looking at international benchmarks of performance, including both costs and customer service measures. Doing the right thing for the wrong reasons is better than not doing anything!

The Realities of the Task Ahead

In Part 2, we touched on some of the facets of service management that make it so difficult to achieve. Building the types of sophisticated systems required to achieve end-to-end automated process flow-through requires highly advanced technology. Early pioneers have proved how difficult and expensive it can be to develop such technology in-house, and every one of the early adopters is driving their computing and equipment suppliers to step up and take over that part of the job. Those just now entering the fray will benefit enormously from the work that has gone before—but only if they are smart enough to listen to others' experiences, to learn from them, and to avoid going down the "gee-whiz" rabbit hole.

Service providers are extremely vulnerable right now—particularly to the kinds of myths propagated by suppliers that are pushing a particular technology or other. A journalist recently asked us if we agreed with a statement made by a service provider representative that "CMIP's competition was no longer SNMP, but CORBA." We responded that achieving service management goals is not about competing technologies but about the proper and sensible use of a mix of technologies that can do the job now and in the future. As we said earlier, new architectures are wonderful and easy to put down on paper. But lean service providers need tools, platforms, applications, and a skills base widely available before they can implement any architecture in real systems. As we've seen, technologies such as object-oriented databases can take a long time to mature to the "industrial strength" required by service providers.

Bernie Maier of Bell Atlantic is responsible for standards associated with operational support systems and has no patience for the suppliers' latest claims. According to Maier,

> When you ask them if they support a certain standard, they say, "Well, not exactly, but we have our own version that does the same thing," and when you point out that a particular technology really isn't yet industrial strength, they say "It's okay—we have our own version that does the job." What it all boils down to is just a lot of new jargon for selling proprietary solutions. And that's not what we want. We can't afford proprietary solutions. It is important for us to have common interfaces based on international OSI/TMN standards. But what service providers need to know is that they can implement a TMN approach now—by demanding standards—and still move to object-oriented technology when it's fully cooked for our industry.

It often seems a dull thing to a technologist to simply follow the pack. But things are moving much too fast, and the whole discipline of service management is far too complex to make going it alone a practical reality. This is one place where the goal should be "follow the industry unless there is a specific reason why that's not good business."

Becoming Better Educated

Service management is nothing if not complex. It's hard to imagine any other area that touches on more aspects of the communications business than service management. So don't expect to become an expert overnight. If you have stayed with us, you should now have a pretty solid background on the business issues and some guidance for charting your direction. But you need much more if you are going to successfully implement service management improvements.

A logical place to go for education is the NMF, which has been in the vanguard of the service management issue. It holds member seminars, conferences, and working sessions on a regular basis to bring its members up to date and explore new and looming problems. Participation in specific teams gives companies and individuals an opportunity to solve specific problems with the help of their peers and suppliers, who are often able to provide more valuable real-life education than any consultant.

In addition to the NMF, participation in other organizations can be helpful in specific areas. X/Open is the industry's place to address the development of a common computing environment, and the Object Management Group is working to define the infrastructure that will implement object-oriented technology. The Desktop Management Task Force is addressing the management of desktop devices and has developed a specific management interface for that purpose. Technology-specific groups, such as the ATM Forum and the SONET Interoperability Forum, address management issues with regard to the devices that implement those technologies.

The various groups of the U.S.-based Alliance for Telecommunications Industry Solutions (ATIS) address implementation of agreements between U.S. local and interexchange carriers, several of which have implemented the OMNI*Point* 1 trouble administration specifications. Eurescom, a European-based consortium of telecommunications companies, has quite a few management-related activities underway, and ETIS (European Telecommunications Informatics Services) focuses on the IT needs of European service providers. There are other groups, as well, involved in testing, standards-setting, and so on.

Applying What You Learn

In Part 4, we have tried to give you some specific steps to follow. Building a strong message and a business case are definite musts. Looking at your internal operations—perhaps using a tool such as SERVQUAL—will help you focus on areas where the biggest gains can be made.

Making the case is becoming easier as time goes by. More companies are able to provide testimonials of their successes in this field and the value of taking advantage of industry agreements in their management systems. Earlier, we cited GTE's experiences in implementing a common trouble ticket with multiple interexchange carriers, specifically the 100-fold decrease in implementation costs between the first and second implementations through the use of common agreements. That's the kind of savings that can make a business case convincing.

Keith Miller, as director of service management at Concert, the joint venture between BT and MCI, noted the speed of implementation that industry agreements can foster. According to Miller, this new company "went from concept to $1 billion in sales in 18 months." Miller cites the early frustrations in not being able to do business effectively across the boundaries. "We used everything we could find to pull this company together. We tried to use industry agreements wherever possible because it meant there was technology we could use. Where agreements existed, we didn't have to waste precious time reinventing everything. Having said that, it will be some time before there are enough vendors with these tools for service providers to use them for the majority of the work of systems integration."

This is not a technology race we are in, but a race to win business. Service providers need to pull out all the stops and take advantage of every possible tool along the way if they are going to make it through the shakeout stage, through the poaching by other industries, and through the internal turmoil that such massive changes inevitably bring. This is the time to make sensible business decisions rather than to show the world that you know the latest technical jargon.

A Final Word

It's not easy, but it's not impossible. In fact, it's a whole lot easier to chart a successful path toward service management excellence today than it was five years ago or even a year ago. What we've presented here is what we've learned and what is available today. But keep in mind that things are changing very fast in this industry. What looks to be an excellent way of getting started on automating service management processes now—with TMN as the base and a whole set of detailed implementation agreements—may be out of fashion tomorrow as new "disruptive" technologies show us better ways.

If you are in this business, and if you are going to make a difference for your company, you simply must begin to steer your organization along a common path. Fragmenting your effectiveness with myriad different approaches will lose you time, cost you a lot of money, and ultimately fail. Make certain that your company's resources are spent solving actual business problems—not just designing a better mouse trap for a world in which mice no longer exist.

And now, we leave you to your task. We look forward to seeing you around the industry, actively helping to keep the communications business financially healthy as it sheds its excess weight and tones its flabby muscles to become again a spirited, fast-moving, and lean child of technology.

Index

Illustrations are in **boldface.**